W9-AEH-511

TS
156
.S362
1998
c.2

ISO 9000 Quality Management System Design

Optimal Design Rules for Documentation, Implementation, and System Effectiveness

Jay J. Schlickman

ISO 9000 Quality Management System Design

Library of Congress Cataloging-in-Publication Data
Schlickman, Jay J., 1934–
ISO 9000 quality management system design : optimal design rules for documentation, implementation, and system effectiveness / Jay J. Schlickman.
p. cm.
Includes bibliographical references and index.
ISBN (invalid) 0-87389-389-0 (alk. paper)
1. Quality control. 2. ISO 9000 Series Standards. I. Title.
TS156.S362 1998 98-12321
658.5'62—dc21 CIP

© 1998 by ASQ Quality Press

All rights reserved. No part of this book may be reproduced in any form or by any means, electronic, mechanical, photocopying, recording, or otherwise, without the prior written permission of the publisher.

10 9 8 7 6 5 4 3 2 1

ISBN 0-87389-389-0

Acquisitions Editor: Roger Holloway
Project Editor: Jeanne W. Bohn

ASQ Mission: To facilitate continuous improvement and increase customer satisfaction by identifying, communicating, and promoting the use of quality principles, concepts, and technologies; and thereby be recognized throughout the world as the leading authority on, and champion for, quality.

Attention: Schools and Corporations
ASQ Quality Press books, audiotapes, videotapes, and software are available at quantity discounts with bulk purchases for business, educational, or instructional use. For information, please contact ASQ Quality Press at 800-248-1946, or write to ASQ Quality Press, P.O. Box 3005, Milwaukee, WI 53201-3005.

For a free copy of the ASQ Quality Press Publications Catalog, including ASQ membership information, call 800-248-1946.

Printed in the United States of America

Printed on acid-free paper

Quality Press
611 East Wisconsin Avenue
Milwaukee, Wisconsin 53202
Call toll free 800-248-1946
http://www.asq.org

Table of Contents

SECTION 3: QMS STRUCTURE

SECTION 4: QMS STYLES

SECTION 5: QMS DESIGN SUMMARY

SECTION 6: CASE STUDY—THE GROWTH DIVISION

List of Figures

List of Tables

PREFACE

This book is dedicated

To my wife Judith

for forty-one years of total quality marriage.

ORIGINS

The ISO 9000 schema has matured to the point at which it contains its own scholarship, mythology, and sibling conflicts. The present transition is to one of big business. And by big business we mean a plethora of International Accreditation Boards, Registrars, Trainers, and Consultants under contract to thousands of global organizations. We have termed this group of entrepreneurs the "ISO 9000 practitioners."[1]

Purpose

The ISO 9000 practitioners work within an exciting and dynamic enterprise that now fosters a myriad pattern of Standards and interpretations of those Standards. It is our hope that this book will make a significant contribution to the clarification of this broad range of perspectives—both for those who wish to create an effective quality management system (QMS) and for those who audit those systems.

It is my privilege, as an independent subcontractor, to work with this group of remarkable talents on both sides of the ISO 9000 street. This situation has afforded me the opportunity to serve as both assessor and auditee within the ISO 9000 certification process. My ISO 9000 experience with over seventy organization, has been extremely positive, and it is my wish to share this unique opportunity with the entire ISO 9000 community.

Thesis

It is our belief that the successful implementation of ISO 9000 in these many and diverse organizations is in part a direct result of a fully compliant and strategically driven QMS.

> ☞ As a result, our design platform consists of a set of design tools that we believe can create a fully compliant QMS whose fabric forms the organization's strategic business declaration.[2]

Readership

In most cases the quality management systems did not come easily. The lessons learned during these experiences should not be lost but should be documented for others to evaluate, and hopefully utilize to create their own effective ISO 9000 QMS. We also feel that much of our discourse will prove useful to those suppliers who have already created systems, but would like to bring their efforts to a new level of effectiveness.

However, the single most difficult aspect in the creation of an effective QMS is the need to address a broad audience. It is also the most difficult aspect of this design approach, and we have chosen to write at a technical level that reveals the operational strength of the International/National Standard. When the term "Standard" is used alone, we refer to the American National Standard ANSI/ISO/ASQC Q9001-1994.

In fact, the more detailed the reader's ISO 9000 background, the better.[3] The conceptual nature of the Standard is not easily envisioned because of its hierarchal nature and descriptive style. We have worked very diligently to clarify and to offer alternative ways to address such issues.

Specifically, the text has been written for a diverse audience:

- Executives who wish to understand what an effective QMS looks like and want to ensure that the system is economically sensible and in concert with the organization's strategic goals
- Members of steering committees, stewards, and ISO 9000 management representatives who must decide on the scope and design detail of the QMS configuration and who must ensure that the system is effectively implemented
- Operational and audit team members who need to understand how to write an effective set of ISO 9000 documents and how to make sure that the system is measured effectively and contains a dynamic corrective and preventive action process
- ISO 9000 practitioners who are interested in the study of self-consistent QMS configurations and what it is like to work on the other side of the table
- Training course suppliers who can use the book as either a research source or as the day-to-day text

Organization

This book establishes a set of design rules for effective QMS creation. In our dialogue, we address the need for full compliance to each "shall" of the Standard and present various system design configurations and strategies.

Section 1

Establishes the overall structure of an effective QMS and the Quality Manual's central role in this exercise. The need for strong organizational leadership is emphasized. Our primary strategy is to establish and maintain a QMS that stresses integrated business and quality management objectives.

The subsequent sections create a series of operational tools for use as the design platform.

Section 2

Establishes the key components of an effective QMS in terms of the Standard's global mandatory requirements. We acknowledge that, effectively, 320 "shalls" are to be clearly addressed within the Quality Manual (we demonstrate how to count them). Special mandatory requirements established at the accreditation board level are also addressed—for example, certification scope and factored items.

Section 3

Deals with the top-down structure of the QMS and the possible methods for document format and linkage. The section creates the framework for effective quality policy statements. Such statements form the Quality Manual's fabric and the glue that holds the entire QMS together as a unified, effective, and affective organizational model. The section also includes sector specific considerations.

Section 4

Considers QMS style and format. Matters of detail, perspective, and affective QMS presentation are considered. The QMS must be both technically accurate and user friendly, and must satisfy a diverse set of readers—the customer, employees, vendors, third-party auditors. Such issues need to be considered at all levels of documentation and the use of specific design configurations—for example, Hub documents, and total business processes—are presented to aid in this effort.

Section 5

Blends all of the tools together and summarizes their use in the creation of a fully compliant and strategically business oriented QMS. These tools are deployed in the case study that follows.

Section 6

Contains the case study that describes the certification of the Growth Division of the Stable Corporation to the Standard. The exercise is based on a wholly fictitious (although you will probably spot yourself), but completely formed, high-tech organization that utilizes our set of design tools. The division chooses a stand-alone form of Quality Manual, that is, a Quality Policy Manual, and the indented text in the study can be removed to form a complete and compliant document.[4]

Join the group and see how the Growth Division breaks through its financial difficulties as they travel the road to continuous improvement. Of course, there is a very wise consultant on board.

Appendices

Are used to present additional detail in regard to tool application.

Conclusion

We believe that adherence to the proposed design rules will create a QMS that is both fully compliant with the ANSI/ISO/ASQC Q9001-1994 and ISO 9001:1994 National and International Standards, respectively, and one that makes a statement about the organization's technical competence, quality commitment, and business uniqueness.

ACKNOWLEDGMENTS

This book is a result of the last five years spent completely immersed in ISO 9000 assessments and consultations, and it effects a blend of thirty-five years of professional experience to create a clearly business-oriented text.

Many organizations and individuals have contributed to the creation of this research, and, with the fear that all writers have that someone will be missed, I wish to thank the following groups who have had an inordinate effect upon my ISO 9000 perspective:

Accreditation and Approval Boards

The ASQ/ANSI Registrar Accreditation Board (RAB); Dutch Accreditation Board (RvA); and the Automotive Industry Action Group (AIAG).

Registrars

Bureau Veritas Quality International (North America), Inc., Jamestown, NY; Scott Quality Systems Registrars, Inc., Wellesley, MA; and TUV America, Danvers, MA.

Consultant Organizations/Trainers

Information Mapping Incorporated (IMI), Waltham, MA, and its International Quality Systems Division (IQS); POWER Incorporated, Salem, MA; Management Software International, Inc., Stoneham, MA; and Corporate Development Services (CDS), Lynnfield, MA.

Direct Clients

SMT East, now USANE, Taumton, MA; and Dome Imaging Systems Incorporated, Waltham, MA.

Special Notice

I would also like to thank the following colleagues specifically for their many contributions, both knowingly and unknowingly, to this book (in no particular order): Dan Morgan, James R. DiNitto, Gary Deines, Hal Greenberg, Jerry Paradis, Steve Gaudreau, Frank Uttaro, Steve Zis, Bill Poliseo, Karl Titus, Stephen S. Keneally, Warren Riddle, Anthony F. Costonis, Ali Dincmen, Joe McCasland, Karen Snyder, John Bader, Mike Hayes, Robert J. Judge, and Cas Makowski.

ASQ Peer Reviewers and Publication Team

ASQ inputs were many and insightful, and they were incorporated into this research. However, any weakness in the text is purely my doing. You can always teach an old dog new tricks, but sometimes the old dog doesn't do good listening (a modern grandchild aphorism).

Endnotes

1. At this writing, there were nearly 20,000 North American registrations to ISO 9000, approximately 15,000 in the United States. A hundred and one countries had adopted ISO 9000, and the worldwide estimate by the end of 1997 was as high as 250,000 sites. These data are reported periodically in *Quality Systems Update,* McGraw-Hill Companies, Fairfax, VA, 1997.
2. Our thesis was previewed in a paper presented at the ASQC Boston Quality Conference, March 21, 1997 (BOSCON). The paper's title was, "ISO 9000 Quality Policy Manual Design."
3. We have assumed that the reader is at the point at which the decision to go ISO has been made and the necessary historical and decision-making background studies in ISO 9000 are complete. An unusually comprehensive source in this regard is the series of publications by the McGraw-Hill Companies, Fairfax, VA, as noted above, especially *The ISO 9000 Handbook,* edited by Robert W. Peach.
4. The choice of configuration is not meant to imply a "best" approach. It simply represents the most common form of Quality Policy Manual that I have found in working with over seventy manuals. The Tier II, III, and IV documentation described is also based on the most common forms of processes, procedures, and forms that I have observed.

EFFECTIVE ISO 9000 QMS IMPERATIVES

"Strategic quality goals are established at the highest company levels and are a part of the companies' business plans. This concept of strategic quality goals is a logical result of the movement to give quality the top priority among the companies' goals."

J. M. Juran, Juran on Quality by Design
The Free Press, 1992

CHAPTER 1
INTRODUCTION

PART 1: THE ISO 9000 QMS PROCESS

Generic

The process used to create an effective Quality Management System (QMS) for ISO 9000 is directly extendable to the creation of any quality management system based on a standard. Examples of such standards include QS-9000 for the automotive industry and FDA/CGMP 820 for medical devices.

In this text we concentrate on the "ANSI/ISO/ASQC Q9001-1994 Quality Systems–model for quality assurance in design, development, production, installation and servicing" (the American National Standard).[1] Throughout our text, the term "Standard" is used to denote this American National Standard.

The process to produce an effective QMS requires

- The analysis of the Standard's requirements—which are stated in terms of "shalls"
- The introduction of an interpretive scheme based on the author's experience and technical background
- A decision on the total effort to be expended to produce the QMS, that is, the degree of responsiveness
- The integration of business strategy with strategic quality management goals

A decision on the degree of responsiveness requires that we agree on what is appropriate for our purposes. Our response to the specific requirements is to be given in terms of quality policy statements, that is, written statements that explain how the organization conforms to the Standard's requirements.[2]

The quality policy statements then drive the form of the entire documentation system, as well as its implementation, and the manner in which we demonstrate the effectiveness of that system.

Thesis

In this exposition we maintain that

- A fully responsive ANSI/ISO/ASQC Q9001-1994 QMS results in a strategic declaration of the organization's quality and technical competence; by fully responsive we mean a QMS that integrates business strategy with quality management and fully conforms to the Standard's "shalls."

- Whereas a paraphrased set of quality policy statements results in a less than effective QMS; by paraphrased we mean a playback of the Standard's descriptive requirements, as opposed to prescriptive statements that indicate the methods used to actually conform to the Standard.

Our Goal

Reality exists somewhere between these two limits. Our goal is to present a set of tools that we believe can produce an ANSI/ISO/ASQC Q9001-1994 QMS that is an effective strategic declaration of the organization's business objectives.

We firmly believe that the intrinsic value of the Standard is its bottom-line focus on productivity and thus profitability—regardless of how the supplier wishes to state such objectives, for example, lowered customer complaints, increased return on investment, lowered rejects, increased repeat purchase orders, and/or lowered product return rates. The Standard, through its inherent continuous improvement paradigm—stress on customer satisfaction, heightened awareness of a lowered cost of quality, transparent business/quality objectives, and explicit calls for process/procedural analysis—offers the supplier a unique opportunity to improve its competitive advantage. However, we see this potential for enhanced marketability dependent upon the supplier's desire to fully comply with the Standard—to write the documented system in a user-friendly manner for a very wide range of readers, to make a total management commitment to this effort, and to establish a QMS that can be maintained in a cost-effective manner.

Tools

To accomplish our goals, we have described a series of design tools that include the

- Integration of business strategy with quality management
- Use of the inherent continuous improvement cycle
- Need for stewardship
- Effective QMS structures
- Adverse effects of paraphrasing
- Use of different media
- Development of quality policy statements
- "Shall" analysis method
- Manual sequence methods
- Possible manual configurations
- Sector specific requirements prescribed by ISO 9000 accreditation boards

We hope that the reader, armed with this set of tools, will create a QMS that represents the true nature of his or her organization and supports its competitive advantage.

Pictorial

The QMS process and our thesis is illustrated in Figure 1.1, entitled, "The Quality Management System (QMS) Thesis."

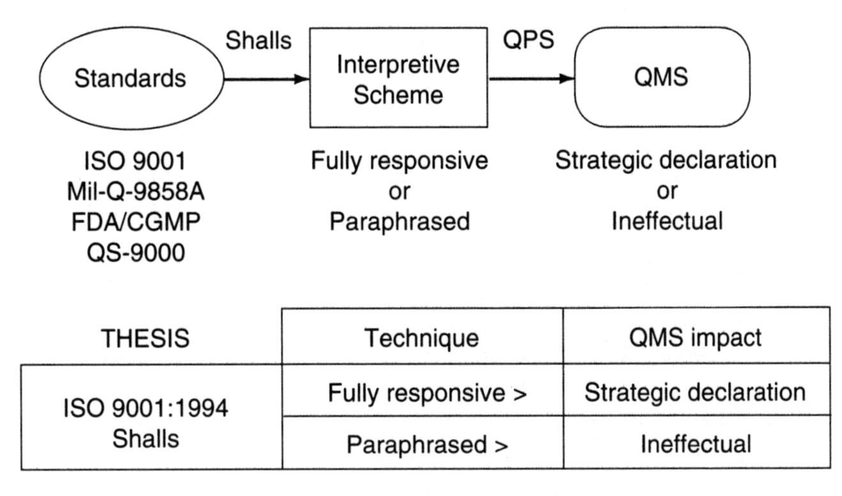

THESIS	Technique	QMS impact
ISO 9001:1994 Shalls	Fully responsive >	Strategic declaration
	Paraphrased >	Ineffectual

Figure 1.1 The quality management system (QMS) thesis (QPS = quality policy statements).

PART 2: THE QUALITY MANUAL CONTROVERSY

The Issue

In our attempt to create a fully responsive QMS, we have found that the *structure* of the Quality Manual (Manual) is the most controversial element in the creation of an effective ISO 9000 documentation system. By structure we mean the manner in which the ISO 9000 elements are discussed in regard to policy; the degree of detail required to describe such policies; the location of policy within the Manual versus location in lower-tier documents; the use of paraphrasing instead of clear, descriptive text; and the choice of presentation as either a stand-alone document or an integrated set of documents.

All of these topics are discussed in detail within this book, and the reader is presented with a number of alternative approaches to the Manual's structure—all of which, we believe, are in full compliance with the Standard.[3] This observation is based on the review of over seventy manuals during five years of firsthand experience in the accredited certification of 50 companies ranging in size from eight to 2000 employees—in industries as diverse as printed wiring assembly contract manufacturing and laboratory-bred animals. (We have already observed the same issues in the areas of QS-9000 and ISO 14000, based on a much smaller firsthand sample.)

This wide interpretive disparity of the ISO 9000 guidelines—especially the interpretation of the ISO 10013:1995 "Guidelines for developing quality manuals"—has proven to be antiproductive for both suppliers and ISO 9000 practitioners. Unfortunately, the clients have ended up in the middle of the conflict and have become captive audiences who must then agonize over what is best for their organizations based on controversial information.

Manual Value

We believe that a clear controversy exists, but that the extent of our efforts in resolving the controversy is not as clear. As a result, we first need to examine the value of the Manual in the ISO 9000 QMS and to clarify its strategic role in both the certification process and the development of an effective quality management system.

Major Gate

As indicated in Figure 1.2, entitled, "Typical ISO 9000 Certification Gates," the Quality Manual is the first critical gate an organization must pass through to complete its quality system. We note that the Manual is derived from an analysis of the organization's total processes—from its strategic business to its service-related activities, from marketing and sales to the repair of returned product. Once the Manual is complete, the process/procedural documents can be finalized and a total quality auditing system can be put in place to monitor the effective implementation of the QMS. By "total quality auditing" we mean audits at the system, process, and product level—both internally and at the subcontractor's facility.

First Seen

In addition, the Manual is the first document seen by the registrar when the time comes to schedule the initial (certification) assessment. Unless the Manual is acceptable, the registrar normally can go no further into the process. Manual review is often accomplished by the registrar as either an off-site activity or part of a documentation review in the facility prior to the on-site initial assessment.

Binder

As we shall indicate throughout this text, the Manual is the major driver for effectiveness in the system and it forms the "glue" that binds all the elements of the system together.

Competitive Advantage

We have observed that a fully compliant Manual—which reflects both the personality and technical competence of the organization—significantly enhances the organization's competitive position. In comparison, we have observed that an inadequate Manual has served as a competitive *disadvantage* and a source of delay in the certification process.[4]

Conclusion

Based on the preceding characteristics, we have concluded that the controversy needs attention.

Observed Causes

We believe that the controversy is a result of rapid growth in the ISO 9000 industry. For example, approximately four to five years ago, in the very early days of

Figure 1.2 Typical ISO 9000 certification gates.

the entry of the United States into the world of ISO 9000, we found that the Manual's structure was essentially a nonissue. We believe that this was due to

- Relatively few suppliers who were large, multidivisional companies with established quality assurance and quality control departments[5]
- Basically high tech organizations
- A limited number of registrars
- Lead assessors with similar quality assurance backgrounds
- A strong total quality management (TQM) influence
- Basic quality programs formed from Mil-Q-9858A and FDA/GMP 820 Standards.

We observed that the Manual controversy grew slowly after 1994, and then in the period 1995 to the present accelerated into what we believe is a major issue. We have concluded that this change is due in part to

- The explosion of candidates in small, medium, and large organizations in extremely diverse fields
- Candidates in widely ranging levels of technology

- A plethora of registrars and consulting groups and more on the way[6]
- A broad spectrum of lead assessors with varied backgrounds
- Enhancements on TQM, for example, reengineering, quality function deployment (QFD)
- The introduction of both integrated Standards, for example, QS-9000, ISO 14000, FDA/CGMPs; and a profusion of ISO 9000 Guidelines

Harmonization

We offer our approach to ISO 9000 QMS design in an attempt to harmonize such widely disparate perspectives. We feel that our thesis forms a common ground for discussion and agreement.

Part 3: Strategic Framework for the Manual

As Integrator

We maintain that the

❑ ISO Driver ❑

☞ Manual should integrate business strategy with quality management.

The Manual becomes a dominant part of the business strategy when it serves as the fabric upon which is imprinted the vision of the organization. The quality policy statements, which form the Manual, drive the operational processes—which in turn form the basis of total quality management. This process forces every author to think deeply about the organization's mission and purpose.

Total Business and Quality Policy

A way in which the business strategy can be integrated within the ANSI/ISO/ASQC Q9001-1994 quality policy requirements is demonstrated in Table 1.1, entitled, "Total Business and Quality Policy Format." In this table we have written the Manual's response to the Standard's element 4.1, Management Responsibility, in the sequence indicated.

When the Manual unifies business and quality strategies into one, it becomes the organization's repository of operational knowledge, which can form the basis of a "learning organization."[7] In addition, it can also be used as the basis for the organization's information technology (IT) imperative that supports process development—the cornerstone of enterprise reinvention, which can result in customer delight.

TABLE 1.1

Total Business and Quality Policy Format

(Suggested response to Section 4.1 of the ANSI/ISO/ASQC Q9001-1994 Standard)

ISO 9001 Clause	Manual's Paragraph Labels and Content	Typical Paragraph Content
4.1	**Business and Certification Scope** (requires registrar's acceptance)	▪ Describes history, products, and locations of the business covered by the certification assessment. ▪ **Example: "The Excellent Corporation designs, manufacturers, and markets SMT assemblies worldwide. All facilities are located in Boston, MA."**
4.1	**Vision Statement**	▪ Defines very long-range business objective. ▪ **Example: "The Excellent Corporation intends to achieve a dominant and globally recognized market position in the SMT industry."**
4.1	**Mission Statement** Can be corporate level and/or divisional/ department level.	▪ Defines key objectives required during the next several years to achieve the Vision. ▪ **Example: "To achieve dominance the Excellent Corporation will expand to new facilities in close proximity to its customers."**
4.1.1	**Quality Policy Statement**	▪ A relatively short thematic statement that embodies the basic quality principles that every employee can remember. ▪ **Example: "Quality within the Excellent Corporation means never being satisfied with anything less than a delighted customer."**
4.1.1	**Quality Objectives (metrics)**	▪ A list of key measurements that are used to define organizational success. ▪ **Example: (a) customer satisfaction (customer returns and complaints, reorders); (b) internal improvements (yields, scrap); and (c) preventive actions.**
4.1.1	**Customer Needs and Expectations**	▪ Discusses how the organization obtains knowledge of customer requirements, satisfaction, and dissatisfaction. ▪ **Example: "The Excellent Corporation formally surveys its customer's satisfaction and dissatisfaction levels quarterly."**
4.1.1	**Propagation**	▪ Discusses how the total quality policy is communicated to all employees, and who is responsible for its implementation and maintenance. ▪ **Example: "The Excellent Corporation holds quarterly meetings with its entire staff to review quality progress and status."**

Proprietary Information

Clearly, proprietary information need not be presented as part of the Manual. For example, marketing strategies, cash flow, and profit and loss information are readily placed in a separate business plan, which is then referenced in the Manual.

Any situation in which proprietary information could become an issue should be discussed ahead of time with the registrar. Usually, third-party lead assessors are very cautious about reviewing either proprietary business information or personal employee data.

ISO 9000-1 Supports a Unified Approach

ISO 9000-1:1994, entitled, "Quality management and quality assurance standards—Part 1: Guidelines for selection and use," is one of a number of ISO 9000 support documents. Part of it is summarized in Figure 1.3, entitled "A Portion of the ISO 9000 Documentation Structure" for the ISO series.

The ISO 9000-1 document is shown on far left in the figure. The ISO 9001:1994 contractual Standard is shown approximately in the center of the page in large format and the ISO 10013 quality manual guideline is shown on the far right at the bottom of that column.[8]

Accreditation Impact

Contrary to common belief, guideline documents are sometimes specified by the accreditation boards—via the Registrars—as strict requirements for certification. Some typical examples are

- ISO 10011-1:1990, Guidelines for auditing quality systems—Part 1: Auditing.
- ISO 10011-2:1991, Guidelines for auditing quality systems—Part 2: Qualification criteria for quality systems auditors.
- ISO 10011-3:1991, Guidelines for auditing quality systems—Part 3: Management of audit programmes.
- EN 45012:September 95, General criteria for certification bodies operating quality system certification. For example, clause 18 requires the supplier to keep a record of all customer complaints and corrective actions taken in regard to such complaints. EN stands for "European Normal." The EN series consists of many supplementary ISO documents.

We suggest that ISO 9000-1 clearly supports our contention that anything less than this unified approach short-changes the purpose of the Standard.

This philosophy is stated in ISO 9000-1:1994(E), Par. 4.2

➪ "Every organization as a supplier has *five principal groups of stakeholders:* its customers, its employees, its owners, its subsuppliers and society.

➪ The supplier should address the *expectations and needs of all its stakeholders . . .*

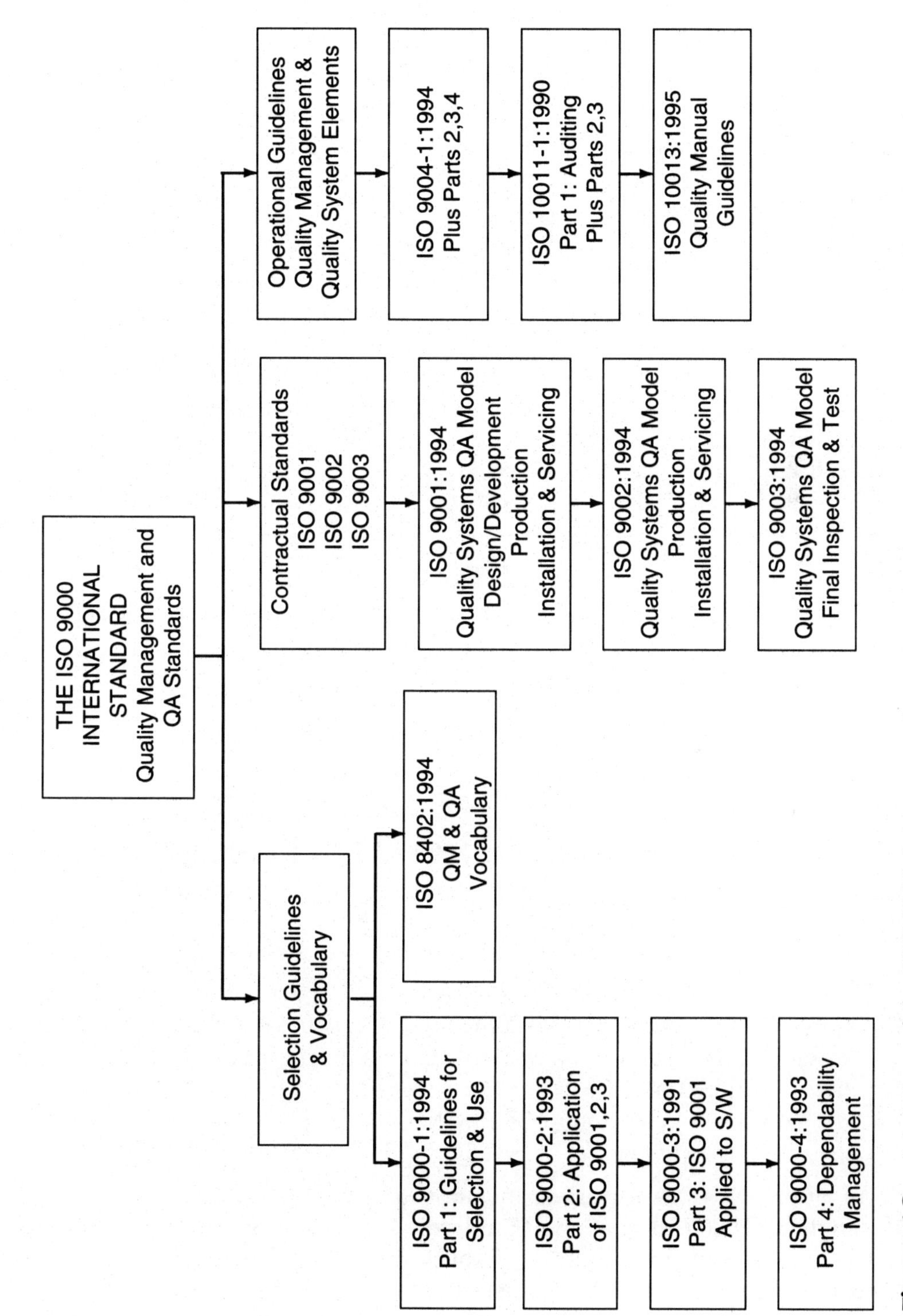

Figure 1.3 A portion of the ISO 9000 documentation structure.

Supplier's Stakeholders	**Typical Expectations or Needs**
➪ Customers	➪ Product quality
➪ Employees	➪ Career/work satisfaction
➪ Owners	➪ Investment performance
➪ Subsuppliers	➪ Continuing business opportunity
➪ Society	➪ Responsible stewardship

➪ . . . The International Standards in the ISO 9000 family focus their guidance and requirements on satisfying the customer . . . ".

A Manual that begins with element "4.1 Management Responsibility" of the Standard from this unified perspective indicates that the supplier has considered ISO 9000 in terms of its overall business directions, and has carefully determined how the quality management system serves to support these directives.

In fact, we have observed a number of organizations that have chosen this unified approach as the starting point for their Manual. In our opinion it has resulted in a very effective document in terms of its value to decision-making readers. Such suppliers have received very high marks on their Manual's from both customers and registrars.

❑ ISO Driver ❑

☞ Most importantly, a unified business-and-quality policy signals to all employees that the main purpose of the ISO 9000 certification is to improve the effectiveness of the operation, not just achieve certification.

PART 4: THE PURPOSE OF THIS PUBLICATION

Our purpose, therefore, is to define a QMS that

- Unifies the enterprise's economic needs with its quality requirements
- Optimizes the flow of information to a wide range of users
- Maintains full compliance with the International Standard
- Provides a dynamic presentation of the organization's drive toward a meaningful ISO 9000 QMS
- Proposes a resolution to the Quality Manual issue and thereby provides a basis for a less diverse set of practitioner interpretations

Our approach is based primarily upon an interpretation of the directives stated in the ISO 9000 International Standards and their associated guidelines. The interpretation is made in the context of thirty-five years of experience in the management of high-tech research, engineering, marketing and sales, quality, manufacturing, and service organizations.

Endnotes

1. All references to ISO 9000 documentation are based on either the sixth edition of the *ISO Standards Compendium, ISO 9000 Quality Management*, International Organization for Standardization, Geneva, Switzerland (ISBN 92-67-10225-7) or the American National Standard series by the ANSI/ISO/ASQC.
2. There is a difference of opinion on whether or not the Quality Manual requires quality policy statements for each "shall" of the Standard. This remains a fundamental issue in manual structure. Our interpretation is based upon Annex C of the ISO 10013:1995 Guidelines for Developing Quality Manuals. In this Annex, the example given is the set of quality policy statements that form a response to element "4.17 Internal Quality Audits" of the Standard.
3. It is necessary to deal with each issue separately because the controversial issues are of such magnitude. The result is some textual redundancy.
4. That ISO 9000 certification represents a competitive advantage is clearly understood by the IRS; see "Registered Companies Risk IRS Scrutiny," Quality Systems Update, McGraw-Hill Companies, Fairfax, VA, March 1997, pg. 1.
5. This observation is supported in part by data presented on page 20 of *Quality,* January 1997, which notes that, in general, larger companies have been registered longer than smaller companies.
6. The *Registered Company Directory, North America,* IRWIN Professional Publishing, Burr Ridge, IL, November 1996, lists the profiles of seventy-one accredited registrars and five un-accredited.
7. Peter M. Senge, *The Fifth Discipline: The Art and Practice of the Learning Organization,* 1990. Available from ASQ Quality Press, Item P539 PH: (800)-248-1946.
8. In the ISO 9000 schema, it is recommended that the document ISO 9004-1, entitled, "Quality management and quality system elements — Guidelines" be used to design the QMS. The ISO 9001, 2, and 3 series are then used as contractual agreements.

CHAPTER 2
FRAMEWORK

PART 1: CONTINUOUS IMPROVEMENT INHERENT IN ISO 9000

We have established that the QMS should be a blend of business strategy and quality management (integrated QMS)—in full conformance with the Standard. We now wish to create the implementation framework for this approach.

Performance Rate

A key tactic of our integrated QMS is to set the organization's performance improvement rate such that it

- Achieves the organization's objectives
- Is attainable from a marketing sense

We wish to drive the organization's performance at some continuous rate of improvement by means of a clearly defined quality management system (QMS) defined by the market imperatives.

C/I Is Innate

Fortunately, the ability to define a continuous improvement (C/I) QMS is inherent in the ISO 9000 Standard, and the Standard's process orientation is ideally suited for a reengineering program[1]—if you wish to drive the system at a maximum rate.

Customer Driven

The customer orientation of the ISO 9000 Standard has already been discussed in Chapter 1, Part 3, where we noted that "The supplier should address the *expectations and needs of all its stakeholders* . . . ." For example, the requirements for a customer-driven program is defined in clauses:

➪ 4.1.1 Quality Policy—" . . . The quality policy shall be relevant to the supplier's organizational goals and the expectations and needs of its customers" . . . ;

- ➪ 4.3.2 a) Review—" . . . the requirements are adequately defined and documented" . . . ; and
- ➪ 4.4.4 Design Input—" . . . Design input shall take into consideration the results of any contract review activities."

In this manner, we have demonstrated that the Standard provides us with the platform for an integrated QMS since the Standard's orientation is already aimed at an effective customer relationship.

Next, we need to show that the continuous improvement cycle—desired by both ourselves and the customer—is also inherent within the Standard.

Shewhart Cycle

We can demonstrate the inherent continuous improvement properties of the Standard if we indicate the relationship between the twenty (20) ANSI/ISO/ASQC Q9001-1994 elements and the Shewhart Cycle of Plan-Do-Check-Act as indicated in Figure 2.1, entitled "ISO 9001 Continuous Improvement Cycle by Element."[2] In this diagram, we have placed every one of the twenty elements in a related category of the Shewhart Cycle. The exact placement of the elements is subject to conjecture, but what is important here is that there is a clear 1:1 correspondence with the paradigm.[3]

Plan—The elements of

4.1 Management Responsibility
4.2.3 Quality Planning
4.3 Contract Review
4.4.2 Design and Development Planning
4.9 Process Control Planning

provide the framework in which we place our strategic business plans, marketing and sales protocols, performance metrics, and recorded achievements.

Do—The front-end section of 4.2 Quality System and the attendant operational elements of

4.5 Document and Data Control
4.6 Purchasing
4.7 Control of Customer-supplied Product
4.8 Product Identification and Traceability
4.9 Process Control
4.11 Control of Inspection, Measuring, and Test Equipment
4.12 Inspection and Test Status
4.15 Handling, Storage, Packaging, Preservation and Delivery
4.16 Control of Quality Records
4.19 Servicing

establish the implementation protocols.

Check—The elements/clauses of 4.4.6 Design Review, 4.4.7 Design Verification; 4.4.8 Design Validation, 4.10 Inspection & Testing, 4.17 Internal Quality Audits, and 4.20 Statistical Techniques provide the requirements whereby we monitor our progress against quality goals and analyze the effectiveness of the overall system.

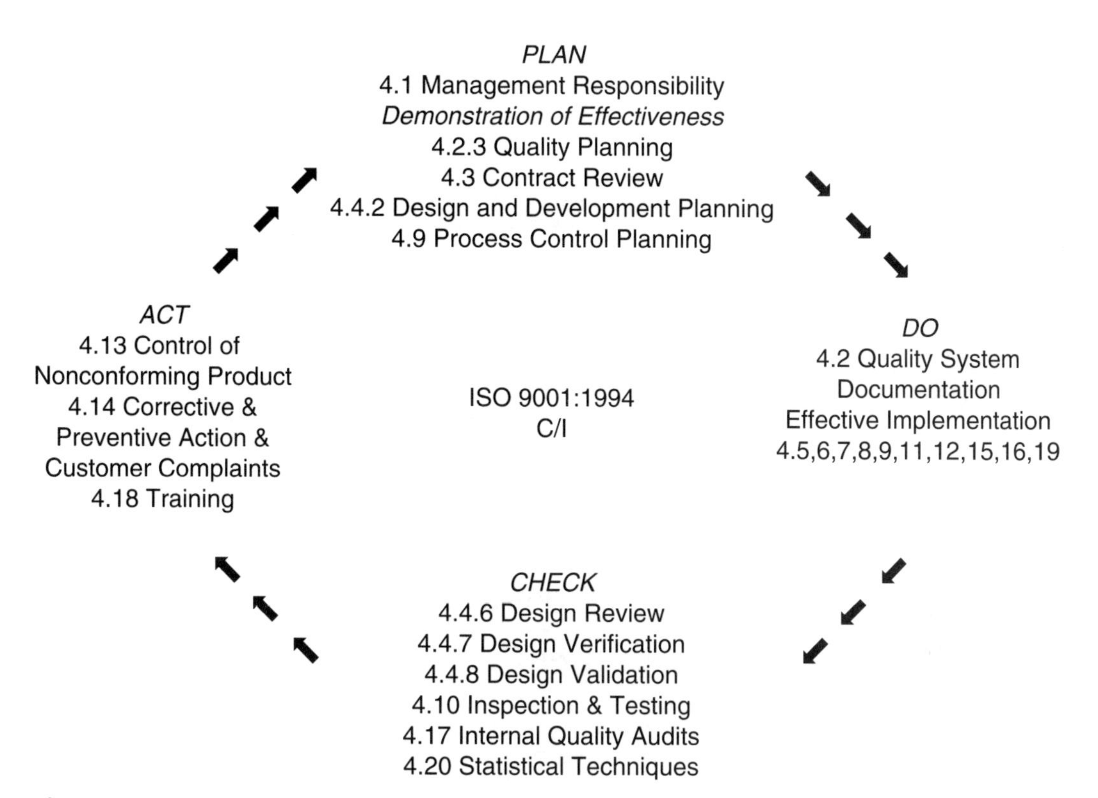

Figure 2.1 ISO 9001 continuous improvement cycle by element.

Act—Elements 4.13 Control of Nonconforming Product; 4.14 Corrective and Preventive Action *with Customer Complaints;* and 4.18 Training establish the methods required to correct those areas that are out of compliance and to establish long-term preventive action programs.

- Thus, when all elements of the cycle are implemented, the paradigm ensures that the system will be documented; that those documents will be used by the employees; and that there will be adequate measurements made to judge whether or not we have demonstrated effective performance against our quality objectives.

Audit Focus

An experienced assessor focuses on elements

- 4.1—especially the Management Reviews, Metrics, and the Preventive Action program
- 4.2—especially the Master Lists of Documents, use of logos, and the currency of Standards and codes
- 4.4—especially on the design review, verification, and validation functions
- 4.14—especially customer complaints and preventive action
- 4.17—especially on the depth of the internal quality audits

not only during the Initial Assessment but at every subsequent surveillance assessment.

This approach ensures that the continuous improvement cycle is maintained throughout the life of the ISO 9000 program. When the Shewhart Cycle is enforced, the odds are very high that the supplier will derive the benefits inherent from an effective QMS.

Assessor Role

Indeed, *the role of the assessor is to teach and clarify*. If this goal is met, the assessor feels fulfilled at the end of a long and intense audit, and the client feels that the effort was worth it. Alternately, if the assessor feels that the goal is to "catch" the client, both parties will end up with a feeling of uselessness.[4]

Part 2: Continuous Improvement Cycle Within Elements

Other C/I

The continuous improvement cycle can also be demonstrated in specific sections of the Standard, for example, Section 4.4, Design Control, as shown in Figure 2.2, entitled "Section 4.4 Design Control—Continuous Improvement Cycle (C/I)."

Further Demonstration

We can further demonstrate that Element 4.9, Process Control—as illustrated in Figure 2.3, entitled "Continuous Improvement Cycle for 4.9 Process Control"—also contains a continuous improvement cycle.

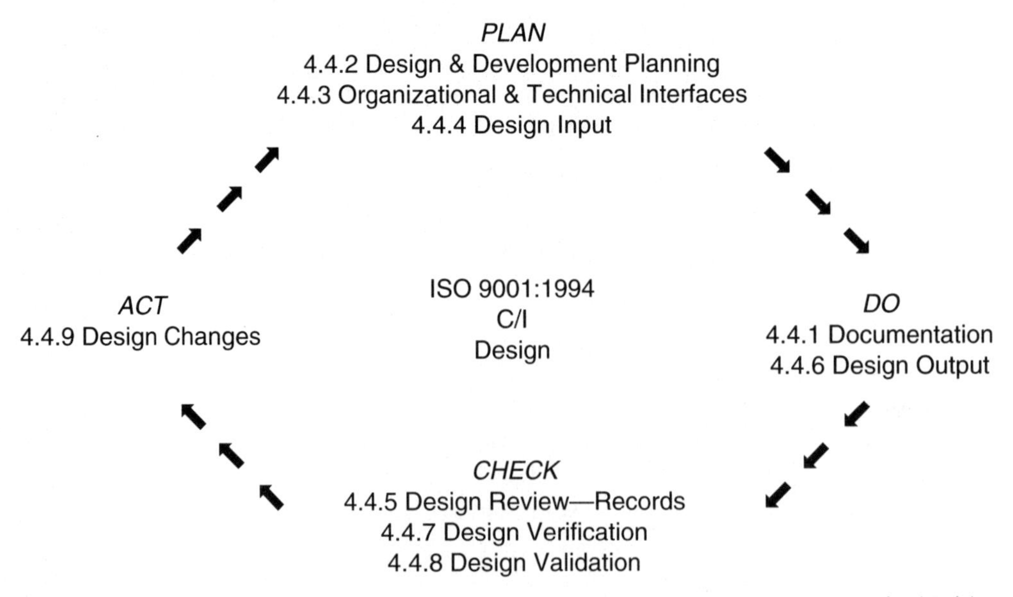

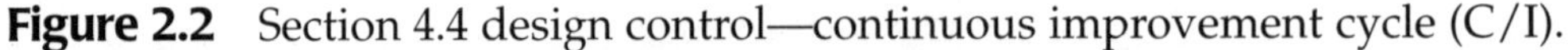

Figure 2.2 Section 4.4 design control—continuous improvement cycle (C/I).

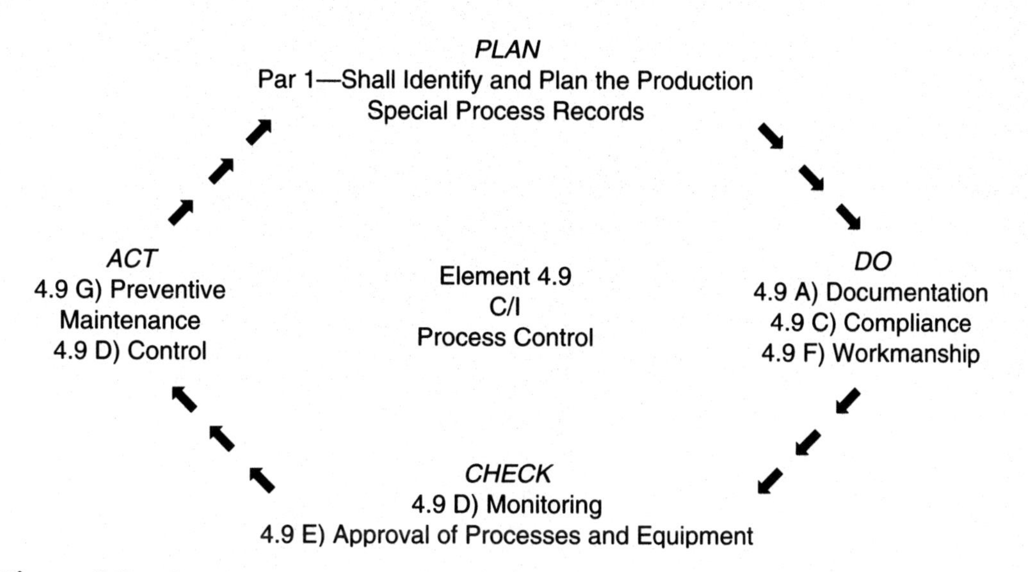

Figure 2.3 Continuous improvement cycle for 4.9 process control.

This phenomenon is a general trend throughout the Standard as demonstrated further in Table 2.1, entitled, "Examples of Other Elements Which Contain the Continuous Improvement Cycle." This table is not meant to be inclusive, but illustrates the general trend. The interested reader will find even more clauses that fit the cycle.

TABLE 2.1

Examples of Other Elements That Contain the Continuous Improvement Cycle

ISO 9001 Element	Plan ➜	Do ➜	Check ➜	Act ➜
4.6 Purchasing	4.6.2 a) Evaluate & select vendors 4.6.2 c) Records	4.6.1 General: Documented procedures	4.6.3 b) Sign off 4.6.4.1 & 2 Verification	4.6.2 b) Vendor control activities
4.14 Corrective & Preventive Action & Customer Complaints	4.14.1 General: Degress of appropriate -ness	4.14.1 General: Documented procedures, implementa-tion	4.14.2 b) Investigation	4.14.2 c) Corrective actions
4.17 Internal Quality Audits	Plan, schedule, & implement	Documented procedures	Audits recorded	Follow-up activities

C/I Conclusion

We conclude that both a market orientation and the Continuous Improvement cycle are inherent within the Standard—whether you wish it or not—and as a result it is necessary to respond to every "shall" to ensure that the Standard's continuous improvement integrity is maintained.

Endnotes

1. The primary role that process analysis takes in the reengineering concept is discussed in *Reengineering the Corporation,* by Michael Hammer and James Champy, Harper Business, 1993, Chapter 8.
2. The role of the Shewhart Control Chart and the development of the continuous improvement cycle is introduced in both Juran on Quality by Design, by J.M. Juran, The Free Press, 1992; and Out of the Crisis, by W. Edwards Deming, MIT Press, 1986. Dr. Deming is very clear on the fact that the "Deming Cycle" is based on the original "Shewhart Cycle."
3. For a lucid discussion of the various applications of the Shewhart Cycle refer to "What Deming Told the Japanese in 1950," *QMJ* 94 Fall issue, p. 9, ASQ Press.
4. The selection of third-party assessors is integral to the selection of a registrar. For a complete exposition on this topic, refer to "How to Select a Registrar," by R.T. "Bud" Weightman, published in *Quality Systems Update,* Special Supplement, August 1996. Weightman is the president of Qualified Specialists, Inc., Houston, TX.

CHAPTER 3
LEADERSHIP

PART 1: ISO 9000 STEWARDSHIP

Clearly, the QMS needs an organizational home. It requires

- Ownership by top management
- A way to be controlled and revised
- Acceptance by all employees

Unfortunately, for most employees, the QMS is something that suddenly appears and overwhelms you with its unaccustomed vocabulary and demands. As a result, it is important to create a documentation system that

- Is worth reading
- Contains phraseology familiar to the industry
- Is relatively easy to work with

For example, we have assumed that the Quality Manual is actually distributed in such a way that all employees can obtain a copy if they so desire. We have found that this is not always the case. We believe it should be, especially since the rest of the documentation system is made available to those employees who need the information on a daily basis.

Full Compliance

Our intent is to make the QMS fully compliant with the Standard's clauses so that we gain the benefit of the inherent interplay whereby one clause either supports or relates to another. In particular, we wish to make the Quality Manual fully compliant with the Standard so that it drives all of the other documentation levels in this direction. We want the Quality Manual to reflect the organization's dedication to an integrated business/quality theme.

Stewardship

We have found that an effective way to ensure that the Quality Manual represents the organization's personality, is distributed appropriately, and is compliant, is to assign various staff members (stewards) with specific sections of the Standard so that they are responsible for the documentation, implementation, and demonstration of effectiveness of each ISO 9000 clause down through all tiers of the system.

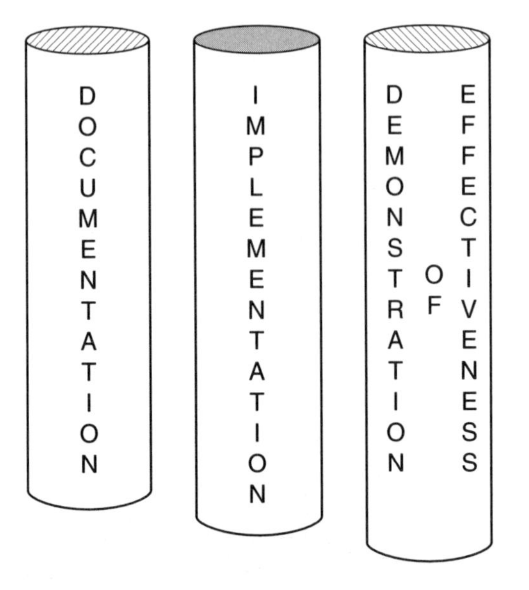

Figure 3.1 The three pillars of ISO 9000.

We will now establish the various roles and duties assignable to the stewards. The requirement for *documentation, implementation, and demonstration of effectiveness*—which forms "The Three Pillars of ISO 9000"—is illustrated in Figure 3.1, and is based directly on clause 4.2.2, Quality System Procedures, of ANSI/ISO/ASQC Q9001-1994(E). This paragraph, 4.2.2 of the Standard, states

"The supplier *shall*

- prepare documented procedures consistent with the requirements of this International Standard and the supplier's stated quality policy, and
- effectively implement the quality system and its documented procedures."

Quality System

A quality system is defined in the ISO 8402:1994 Vocabulary as

- "The organizational structure, responsibilities, procedures, processes, and resources needed to implement quality management."

Quality Management

Quality management is defined in the ISO 8402:1994 Vocabulary as

- "All activities of the overall management function that determine the quality policy, objectives, and responsibilities and implement them by means such as quality planning, quality control, quality assurance, and quality improvement within the quality system."

As a result, if we blend the two definitions, we can graphically demonstrate the functional relationships between the various parts of a Quality Management

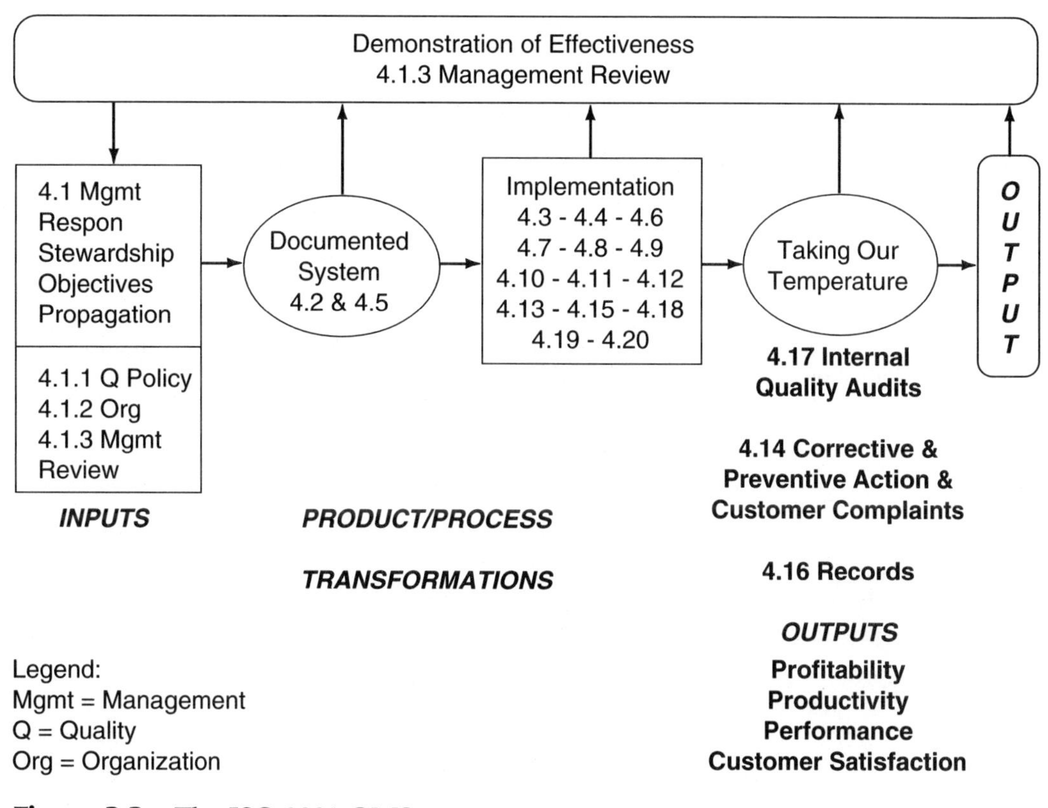

Figure 3.2 The ISO 9001 QMS.

System (QMS). This concept is shown in Figure 3.2, entitled, "The ISO 9001 Quality Management System."

We see in Figure 3.2 that the Standard has essentially defined a true engineering system complete with inputs, outputs, and feedback loops. The inputs of quality objectives as well as quality management protocols are transformed by the system to produce continuously improved processes and products, which lead to outputs that include enhanced productivity, profitability, performance, and customer satisfaction.

The gain of the system can be described quantitatively to some degree prior to certification by measuring how close we are to completeness. A gain of near unity, that is, when we reach 90 percent of our goals, can be used to successfully determine when it is time for the initial assessment by a registrar.

After certification, the gain is based upon how well we achieve our quality objectives, which are measured according to metrics that fit our industry.

The Stewards Take Our Temperature

The key to a successful QMS—one that helps us achieve our quality goals and objectives—is in the ability to measure either our progress or lack thereof. Someone has to be responsible for this task.

A useful chart for this purpose is shown in Table 3.1, entitled, "Taking Excellent's Temperature—ISO 9001 Readiness Chart." In this table, we use the Shewhart Cycle of Plan-Do-Check-Act to create a quantitative matrix for this

TABLE 3.1

Taking Excellent's Temperature—ISO 9001 Readiness Chart

(90% implies readiness for the initial assessment)

C/I Elements	**Relative %**									
Month	**Mar**	**. . .**	**Ju**	**Jul**	**Au**	**Se**	**Oc**	**No**	**Dec**	**Where pts will come from**
Plan—Executive										
Management review	30			***50***	50	***60***				–2 Mgmt Rev(15)
Quality Manual	5			***70***	***80***	***85***				–Ready audit
Objectives—metrics	30			***50***	***60***	***70***				–Quantitative
Demo of effectiveness	5			***50***	***55***	***60***				–Full C&PA&CC
Do—Operational										
Tier II docs	5			***60***	65	***75***				–Final audit
Tier III docs	30			***40***	50	***55***				–In-process
Tier IV docs	30			***50***	60	***70***				–In-process
Implementation	30			***40***	50	***55***				–Start using
Master records list(s)	5			30	50	***70***				–In–process
Master docs list(s)	5			30	50	***60***				–In–process
AVL list(s)	5			30	70	***75***				–Tracking data
Master cal list(s)	5			30	50	50				–In-process
Training program	30			***40***	50	***55***				–In-process
Check—Internal										
Internal Quality Audits	5			30	50	***60***				–5 pts each
Verification	30			***50***	80	***85***				–Ready audit
Validation	5			***50***	60	***65***				–Ready audit
Act—Effectiveness										
Corrective actions	5			***40***	***50***	***65***				–Analyze
Preventive actions	5			***40***	***60***	60				–Make a list
Customer complaints	30			***50***	***60***	***65***				–Analyze

Positive Monthly Changes are highlighted in ***bold and italics***.

Guide
Startup required = 5%
A clear base exists = 30%
Mid-game to ISO readiness = 35–70%
End-game, the last few yards = 75–85%
Ready for initial assessment >= 90%

Legend
AVL = Approved vendor list

Cal = Calibration
Demo = Demonstration

measurement purpose. The paradigm has been given a parallel terminology—Executive-Operational-Internal-Effectiveness—to emphasize the role that is to be played by management.

In the example, the Excellent Corporation, which started from a fairly weak base, has progressed month by month to the point at which the Manual is about one more short pass away from a score, and some exceptional work has been done in the verification of product.

This scorecard is kept up to date weekly by the ISO 9000 management representative, based on inputs from the stewards. Thus we have defined the first key duty of a steward.

Overall Role of the Steward[1]

To further determine the duties of stewardship, we note that in clause 4.1.2.1 of the Standard—Responsibility and Authority the essence of ISO 9000 is to state with great clarity

- who manages
- who performs, and
- who verifies

these activities and functions.

- Thus, the overall role of the steward is to ensure that the quality system, as defined, is measured against goals, and that the functions of management, operational performance, and verification & validation are effectively implemented.

Summary of the Steward's Specific Duties

Specifically, each steward ensures that in his or her channels of information

- There is a *technical correlation* from quality policy down to the lowest documentation levels, e.g. forms—the "channel."
- There is an affective link between each quality policy and the lower-tier documents.
- The entire documentation system is complete.
- The documented system is completely implemented.
- The "channel" is helping to achieve our quality objectives.
- The continuous improvement programs are effective.
- There is an acceptance at all levels of the QMS directives.
- The quality policies are managed at all levels.

Team Leaders

It is also customary to assign cross-functional teams to handle the more broad-reaching requirements that touch several operational areas. The leader for such a group is sometimes called the team leader and for more complex activities can hold the title of program manager. Their responsibility is to ensure that specific cross-functional programs are effectively managed. Such programs include

- ISO 9000 Management Representative (leads Document & Data Control & Records Program and stewards element "4.2 Quality System")
- Corrective and Preventive Action and Customer Complaints Manager (usually the QA manager is the function exists, and if it does not it is usually shared by several managers, e.g., the customer service manager stewards the customer complaints activities)
- Total Quality Audits Manager (again usually lead by the QA manager if the function exists, otherwise stewarded by another manager, e.g., controller, head of operations, director of safety)
- Training Manager (headed up by the human resources manager if that function exists, and, if not, it is usually shared by all local area managers, e.g., all department heads are responsible for the training and documentation of their staffs)

These assignments can take many forms, but a possible distribution of responsibility in a manufacturing facility could be as follows (refer to Table 3.2, entitled "Possible Stewardship Distribution by ISO 9000 Element in a Manufacturing Facility").

An effort has been made in the chart to distribute the stewardship responsibilities evenly among top management. Unfortunately, it often happens that Operations and Quality Assurance end up with an inordinate level of activity compared to the other departments. This type of situation is to be avoided when possible since everyone in the company is usually already overloaded.

Appendix A, entitled "ISO 9000 Stewardship Summary," is a convenient way to catalog both the stewardship and team leader/program manager assignments.

Part 2: Cross-Functional Teams

Section Experts

Most importantly, to produce an effective Quality Manual to drive the entire QMS—one that all employees can understand and relate to—requires that the stewards and authors have a detailed knowledge of the business and that each section is written by an author who has the most operational experience within that area—for example, a technical or operational expert.

It is one of the duties of the stewards to ensure that technical competence is optimized in this regard.

Ineffectiveness

Invariably, ineffective Quality Manuals result from the work of authors who do not take enough time to truly understand the organization's processes. This analysis of process is not only paramount in ISO 9000 but is just as critical in related activities such as reengineering and total quality management.

We have often observed incomplete sections created by someone who is under a heavy time constraint and who is only remotely familiar with the section's

TABLE 3.2

Possible Stewardship Distribution by ISO 9000 Element in a Manufacturing Facility

Position Within Organization	Responsible for Elements
Site Manager	4.1 Management responsibility
ISO Management Rep*	4.2 Quality system 4.5 Document & data control 4.16 Control of quality records
Sales & Marketing Manager	4.3 Contract review 4.19 Service
Chief Engineer	4.4 Design control 4.20 Statistical techniques
Purchasing Manager	4.6 Purchasing 4.7 Control of customer-supplied product
Manufacturing Manager	4.8 Product identification & traceability 4.9 Process control 4.15 Handling, storage, packaging, preservation, and delivery
Quality Assurance Manager	4.10 Inspection and testing 4.12 Inspection and test status
Plant Engineer	4.11 Control of inspection, measuring, and test equipment
Corrective Action Team	4.13 Control of nonconforming product 4.14 Corrective and preventive action w/customer complaints
Internal Audit Team	4.17 Internal quality audits
Human Resources	4.18 Training

*The assignment of specific elements to the ISO management representative during the "set-up" phase is not meant to suggest that this function is not responsible for the overall coordination of the program. It simply levels out the writing and editing load.

content. It is the job of the steward to see that such situations are remedied through thoughtful assignment of personnel and a vigorous support system.

It is unproductive to push the ISO schedule to the point that the organization's effectiveness suffers. There is no rush to certification: It is the usefulness of the created QMS that is important. Consider how often seemingly drop-dead certification deadlines are stretched out by those who originally demanded them. In all cases, they come to realize that the integrity of the business always comes first.

The worst that can happen is that it will take a little longer and cost a few more dollars to complete the program.

It Is Impossible to Fail Certification Unless You Give Up

This is also true for the initial assessment. One does not fail a third-party assessment; this is a part of the ISO mythology. One does get nonconformances that need to be corrected. The worst case is a major finding that could delay the certification process by up to three months and require the additional expense of a return visit by the registrar's lead assessor to clear the nonconformance. But that is it.

The steward's task is to make sure that there are no major findings possible. Inevitably there will be minor findings at the initial assessment, the first surveillance, the second surveillance . . ., the recertification assessment, and the re-recertification Assessment. That is what continuous improvement is all about. By major findings we mean, for example, an ineffectual management review, a poorly managed training program, a lack of internal quality audits, a corrective and preventive action program that is uncertain and loosely managed. The stewards must pay close attention to these areas.

By Teams

As we noted previously in Table 3.2, the elements of the Standard can be grouped together in a way that makes sense in cross-functional team arrangements.

There is nothing mystical about this approach, and each organization should structure its teams in a way that best fits its purpose, but we have found it helpful when design and manufacturing teams form as shown in Table 3.3—Team Member Grouping.

TABLE 3.3

Team Member Grouping

Team Members	ANSI/ISO/ASQC Q9001-1994 Elements
▪ Site manager, sales & marketing & service managers	4.1, 4.3, & 4.19
▪ ISO management rep, document control, finance	4.2, 4.5, & 4.16
▪ Engineering mngrs, program mngrs, mnfg eng'g	4.4, & 4.20
▪ Purchasing mngrs, production control	4.6, & 4.7
▪ Manufacturing mngrs, shipping/receiving	4.8, 4.9, & 4.15
▪ Quality assurance (QA), manufacturing mngrs	4.10, 4.11, & 4.12
▪ ISO mgmt rep, manufacturing engineers, QA	4.13, & 4.14
▪ Audit team mngr, auditors, QA	4.17
▪ Human resource mngr, dept. managers	4.18

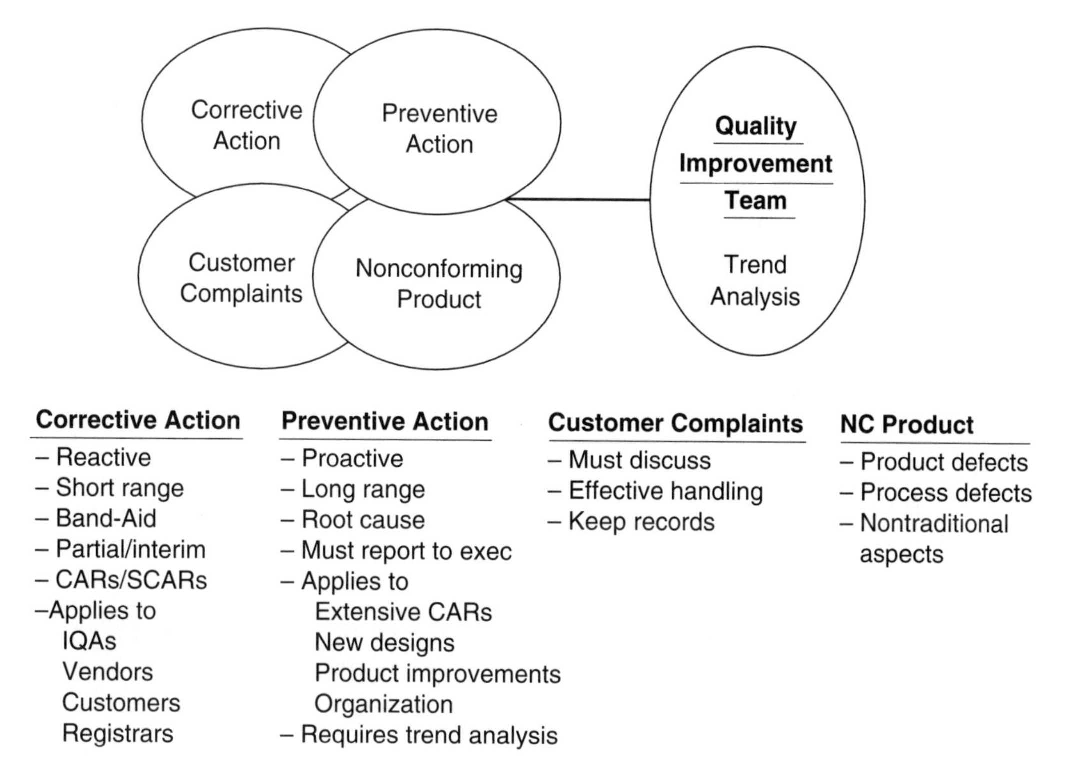

Figure 3.3 Corrective and preventive action w/customer complaints and interface to 4.13 control of nonconforming product.

w/o Design/QC

For those ISO 9002 organizations without a defined quality assurance department, elements 4.10, 4.11, 4.12, and 4.20 would still be covered by manufacturing.

Team Effectiveness

Figure 3.3, entitled "Corrective & Preventive Action w/Customer Complaints & Interface to 4.13 Control of Nonconforming Product," illustrates graphically the need for cross-functional teams in the effective implementation of the Standard. In this case, the team consists of members from QA, engineering, manufacturing, customer service, and finance. The several functions required for each key activity are also listed in Figure 3.3.

We have found that this area requires the most intensive training and takes the longest to optimize. When one looks in detail at the requirements of element 4.13 and 4.14, the difficulty in interpretation is not such a surprise.

We offer some suggestions in this regard:

- Use supplier (vendor) corrective actions reports (SCARs) to manage the interface with subsuppliers (subcontractors).
- Limit 4.13, Control of Nonconfoming Product, to nonconformance reports (NCRs) that occur after incoming (receiving) and prior to shipment.
- Limit corrective action reports (CARs) to the internal quality audit findings and for big-time nonconformances that require a team. CARs are by their nature expensive and time-consuming.

- Run the preventive action program via memos and reports. Stay away from a specific format: Preventive actions, by their nature, are broad in scope and need a lot of creativity to carry them through to completion.
- Manage customer complaints via the returned material authorization (RMA) transactions, on logs kept by sales, and on logs maintained by customer service or an equivalent organization.
- Assign one person to decide on CAR responses, for example, the manager of QA.
- Assign one person to filter the customer complaints for the required response, for example, manager of sales & marketing.
- Assign one person to take primary responsibility for the overall review and trend analysis of the above data bases, for example, the manager of QA. That person prepares and reports to the quality improvement team on the trend data.

Endnotes

1. The use of clause 4.1.2.1 to describe the steward's role is not meant to preclude the total coverage of the clause over all of the organization.

QMS Documentation Requirements

"Taking on too much at once can sap any amount of energy and thwart the successful completion of any undertaking. Parceling tasks into manageable portions without losing sight of the overall design of the whole endeavor is one of the arts of leadership at all levels, from personal self-management to corporate, community, and political domains of action."

The Lost Art of War, *by Sun Tzu II, commentary by Thomas Cleary, Harper San Francisco, 1996, p. 101.*

Chapter 4
Global Mandatory Documentation Requirements

Part 1: ANSI/ISO/ASQC Q9001-1994 Global Mandatory Requirements

Introduction

In Section 1, we indicated our desire to create a business-oriented QMS in which the financial and quality objectives were transparent. We then stated the importance of a powerful organizational team to fully realize the Standard's inherent continuous improvement capability.

We now wish to establish the key components of an effective QMS in terms of the Standard's mandatory requirements. In this regard, we acknowledge that all requirements ("shalls") are to be addressed—the obvious case—but we focus our attention on just the global requirements.

Mandatory Tiers

The so-called ISO 9000 Tiers (hierarchal levels of information) are merely guidelines that have grown out of the military requirements and have become de facto Standard's because of their usefulness. However, we will now demonstrate that only the Quality Manual, procedures, and records are mandatory hierarchal documents in ANSI/ISO/ASQC Q9001-1994.

Of the three, records are the least understood in regard to their position in the documentation hierarchy. In fact, records, that is, historical documents or documents used as objective evidence of activity, are distributed across the elements and can occur at any level of the hierarchy. For example, management review minutes are at the Tier I level and routers/travelers in the form of medical device history formats are at the Tier II level.

We can demonstrate these concepts if we carefully examine the ANSI/ISO/ASQC Q9001-1994 system requirements.

Global Mandatory System Requirements

The requirement for a Quality Management System is denoted in Par 4.2.1 General of the Standard:

➪ "The supplier *shall* establish, document, and maintain a *quality system* as a means of ensuring that product conforms to specified requirements."

The ISO 8402:1994 vocabulary defines a quality system as an

➪ "Organizational structure, procedures, processes and resources needed to implement quality management."

Quality System

There is confusion over just what is actually included in the quality system. For example, we have found marketing & sales and finance ignored in some systems. To our surprise, we also found the computerized information systems barely discussed.

When we point out that marketing & sales is an integral part of customer satisfaction and dissatisfaction, and that finance is an integral part of the cost of quality, the matter is often resolved. The issue with computer systems is somewhat obvious, and once it is pointed out it is usually quickly resolved.

However, there are still discussions over certain parts of the organization that should or should not be left out of the certification. For example, if the decision is made to exclude the maintenance department, make sure that they are not responsible for activities like either the weekly ESD floor covering to prevent electrostatic damage to printed wiring assemblies or the preventive maintenance process that keeps all the paper-making machines rolling.

The crux of this matter is that it is best to include everyone in the organization. We have yet to find any employee in a given organization who did not participate in some way in the quality activities.

Information Flow

In a more graphical sense, Figure 4.1, entitled "Potential Quality System Information Channels," indicates the several possible channels of information that are covered under the ISO 9000 quality system concept.[1]

In practice, it is common to find the QMS defined primarily by the quality management documents channel and the engineering information systems and

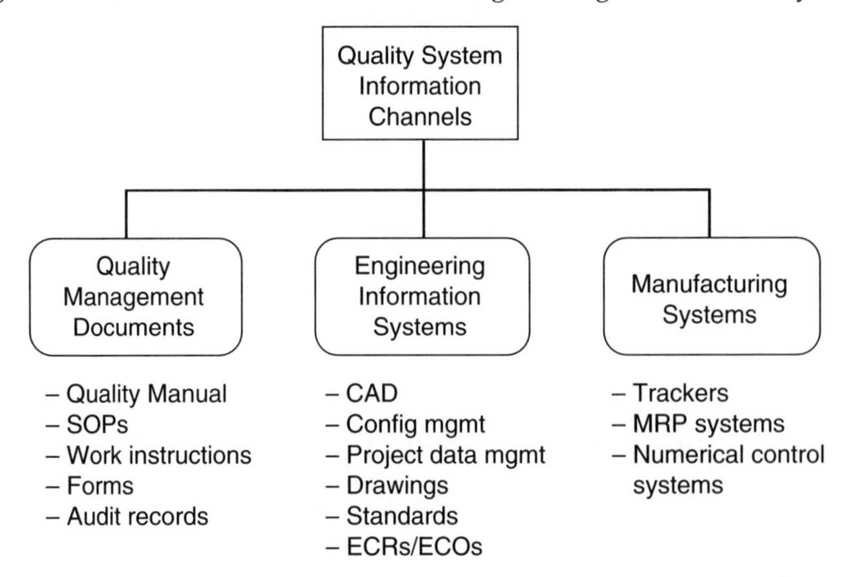

Figure 4.1 Potential quality system information channels.

manufacturing systems channels weakly described. The day-to-day interplay between engineering, operations, and quality assurance requires that all channels be equally efficient.

This is why it is inefficient to certify a design and manufacturing facility to ISO 9002 and then to plan a future time to complete the certification to ISO 9001. The number of daily interfaces with engineering requires interface procedures. It involves essentially the same effort to simply qualify the entire facility to ISO 9001 to begin with rather than creating all of those interface documents.

Global Mandatory Organizational Requirements

It is important to define the employee certification scope via an organizational chart or an equivalent table. A chart is generally used and includes a box for the ISO management representative who normally reports directly to the site manager for ISO 9000 purposes. The Standard requires information as follows:

➪ 4.1.2.1 Responsibility & Authority— ". . . the interrelation of personnel who manage, perform, and verify work affecting quality shall be defined and documented. . . . "

A typical organization chart is shown in Figure 4.2, entitled "The Excellent Corporation's Organization." Notice that *all levels* of the organization are defined. Quite often, the charts are found as an appendix to the main body of the Manual. The actual names of employees are not required, although assessors are very grateful to have such a chart available in addition to the generic one.

It is also important to include a paragraph description of the duties and responsibilities for each of the top managers. This could be an appendix but is normally placed in the body of "4.1 Management Responsibility."

A typical paragraph would read

- Engineering Manager Responsibilities:
  - "Assist M&S in sales of new products.
  - Obtain and allocate engineering resources.
  - Serve as the chief engineer.
  - Schedule and supervise engineering projects.
  - Ensure compliance to ISO 9001 design requirements.
  - Interface with customers to determine design satisfaction."

We also see in Figure 4.2 that the various activities defined in paragraph 4.1.2.1 of the Standard are included in terms of a code in each appropriate box. The code is explained in Figure 4.3, entitled "Responsibility & Authority Legend for Excellent's Organization." We have found a strong area of contention as to why this requires a response. The letter "V" denotes personal verification of your work.

As a point of clarification, the wording in the Standard requires this form of response, although a table would serve just as well:

➪ 4.1.2.2 Responsibility and authority—" . . . particularly for personnel who need the organizational freedom and authority to:

a) initiate action to prevent occurrence of any nonconformities relating to product, process, and quality system;
b) identify and record any problems relating to the product, process, and quality system;

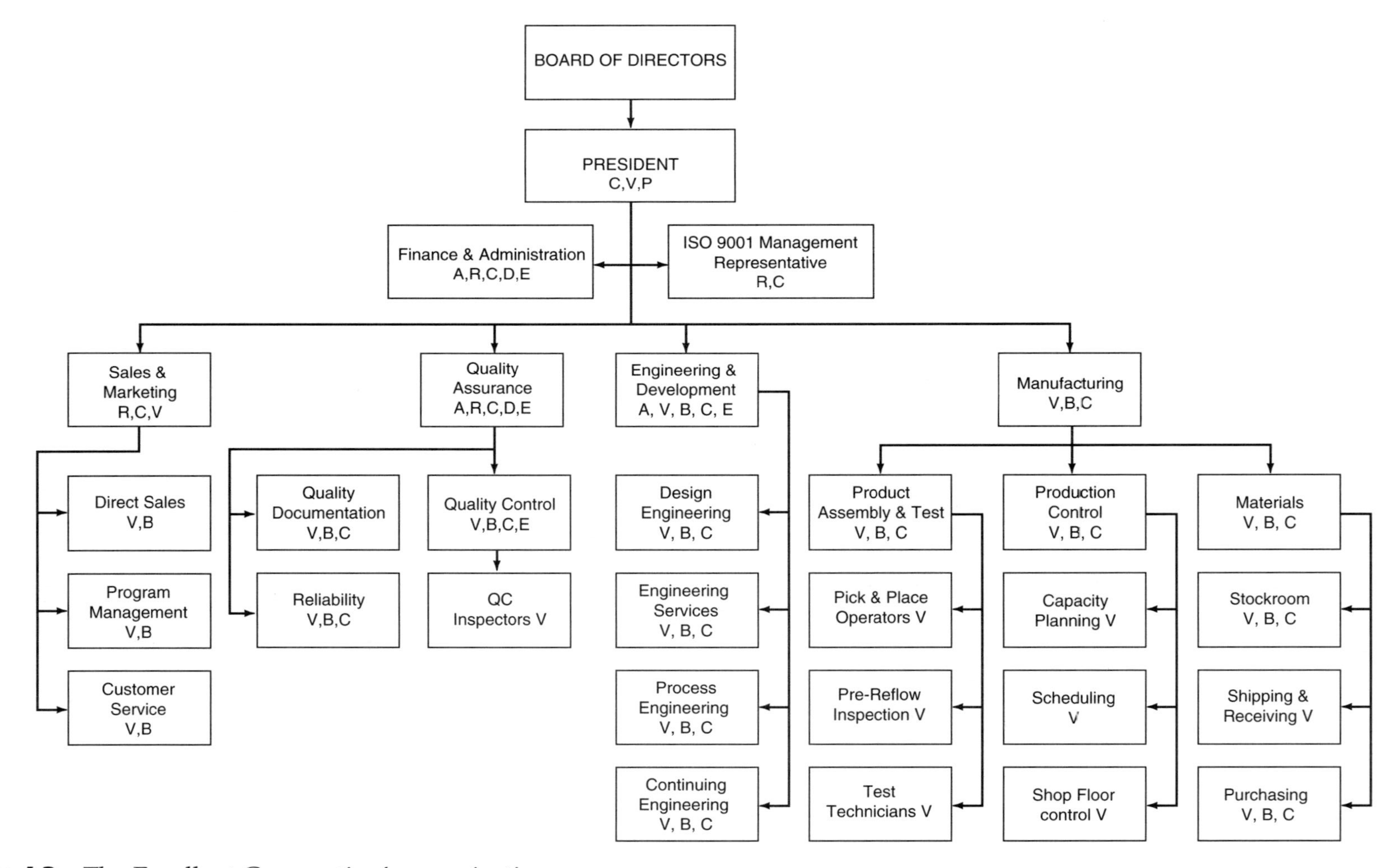

Figure 4.2 The Excellent Corporation's organization.

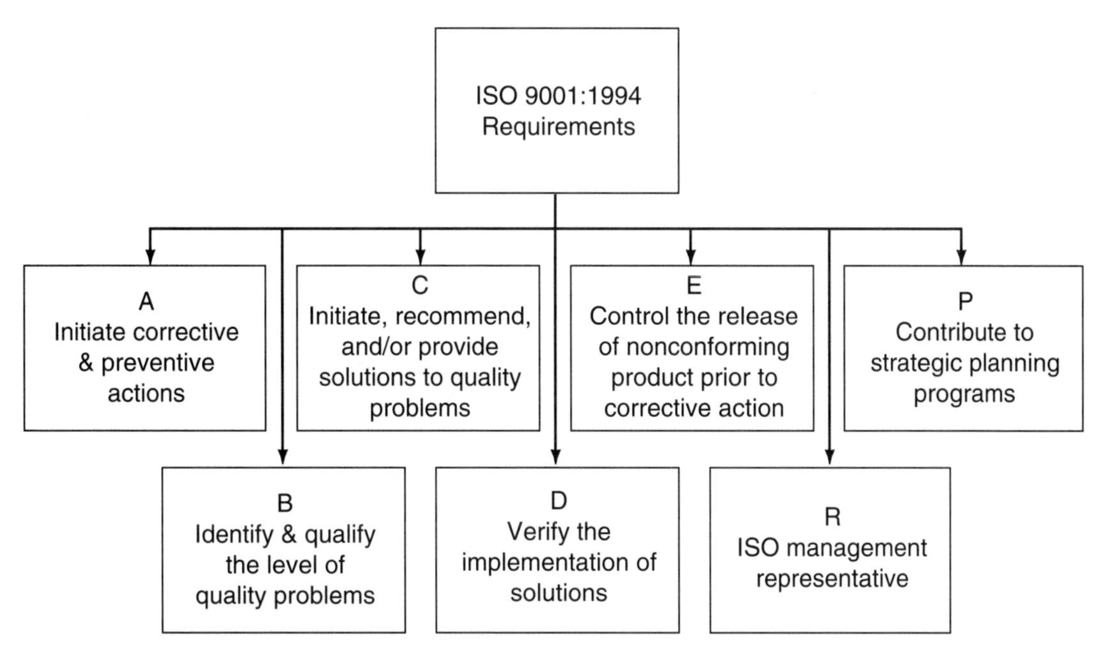

Figure 4.3 Responsibility and authority legend for Excellent's organization.

c) initiate, recommend, or provide solutions through designated channels;
e) verify the implementation of solutions;
f) control further processing, delivery, or installation of nonconforming product until the deficiency or unsatisfactory condition has been corrected."

Job Descriptions

Job descriptions can also be used to define these functions, and specific steps in procedures can be used to enhance the descriptions. However, the Manual must respond to this "shall" and be clear as to the method used to define the activities.

Global Mandatory Interface Issues

In many cases, the certification site is part of a much larger organization. It is then necessary to define the interfaces that exist between the corporate offices and interdivisional sites. This somewhat obscure requirement is extremely important in multidivisional organizations that share operational areas, for example, engineering, purchasing, metrology, and receiving & shipping.[2]

A typical example of such an interface chart is shown in Figure 4.4, entitled "Excellent Corporation's Interdivisional Relationships." As indicated, there are a number of corporate and divisional interfaces. The same information could be demonstrated by means of a table. In fact, when there are several dozen interfaces, a table is easier to understand.

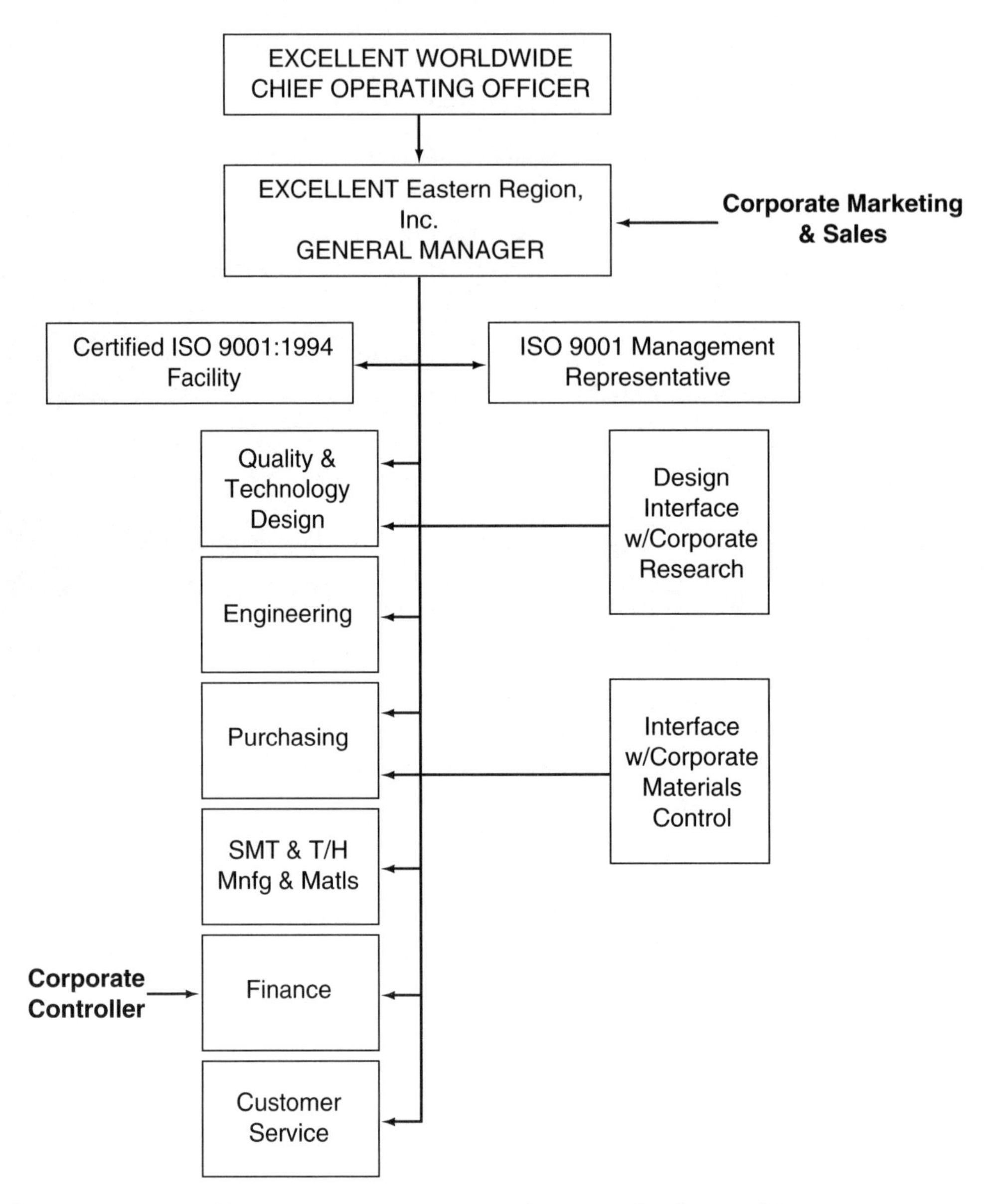

Figure 4.4 Excellent Corporation's interdivisional relationships.

What we see in Figure 4.4 is that corporate marketing & sales provides market analysis and customer leads to the general manager. The general manager works with the in-house customer service department to close and follow up on sales. In addition, the local finance department interfaces with the corporate controller for capital equipment and fiscal budget planning.

Also, the local design capability is enhanced via the research and development facilities at corporate, and local purchasing obtains better price discounts by buying raw material through the corporate materials department.

In many situations the purchase of material from another division is by means of the same purchase order that is used to buy from any subcontractor. Also, when product is shipped from one division to another, the same sales orders are used as in any shipment to a customer. In that case, it is usually acceptable to state in the Manual in response to 4.2 Quality System that

- All buy and sell transactions between the Excellent Corporation's operating divisions are by means of either purchase orders or sales orders which are the same notes used with either vendors or customers.

Otherwise a fairly detailed interdivisional agreement would need to be created—usually in the form of a procedure—and signed off by the two interfaced general managers. Be sure to pass such decisions by your registrar for acceptance and comments.

Mandatory Quality Policy Manual (Tier I) Requirements

The requirement for a Quality Manual is also specified in Par 4.2.1 General,

➪ "The supplier shall prepare a *quality manual* covering the requirements of this International Standard."

Tier II Linkage Requirements

Linkage is also defined in Par 4.2.1 General,

➪ "The quality manual *shall* include or make reference to the quality-system procedures and outline the structure of the documentation used in the quality system.

➪ *Note 6 —Guidance on quality manuals is given in ISO 10013:1993."*

We note the Tiers should be clearly linked so that it is possible to readily navigate throughout the documentation. Section 3, Chapter 9 covers this topic in some detail.

Documented Procedures and an Effective Implementation

Tier II, III

The requirement for procedures and effective implementation is expressed in Par 4.2.2 Quality System Procedures:

➪ "The supplier *shall:*

a) prepare *documented procedures* consistent with the requirements of this International Standard and the supplier's stated quality policy, and
b) *effectively implement* the quality system and its documented procedures."

The measurement of such effectivity is usually via the

- Management review cycle in which the metrics of the organization are analyzed, that is, progress made toward the organization's quality/business objectives
- Corrective & preventive actions and customer complaints programs;
- Internal quality audits
- Design reviews

Mandatory Sensible Requirements[3]

The charge to the authors to create a reasonable volume of documents and to keep the corporate economics in mind is expressed in Par 4.2.2, of the Standard, where the organizational complexity, methods, skills, and training are to be considered in the creation of the quality system documentation. At issue here is the tendency to "overwrite"—usually a good 40 percent more than is necessary for an effective presentation. We all do it, and it usually takes a few years after certification to streamline the system.

There will be time available after the ISO 9000 certificate hangs on the wall to clean up the documentation. Besides, the documents will most likely change significantly during the first surveillance period, which is a good time to make the necessary revisions.

Part 2: Records Mandatory Requirements

The requirement for records is found in element 4.16 Control of Quality Records:

➪ "Quality records *shall* be maintained to demonstrate conformance to specified requirements and the effective operation of the quality system."

This area tends to be one of mass confusion due to a lack of specificity.[4]

Records (used as objective evidence of activities) complement the hierarchal documents and can be associated with any Tier. For example, records do not require a separate documentation control numbering system since they are already controlled, either centrally or locally, by date and signature. Imagine how nonproductive it would be to take a form with a control number, F-103-01, fill it in, and then give it another control number, R-103-01, for storage as a record. This is a danger in ISO document and data control interpretation and is certainly not specified in the Standard.

Records as Historical Documents

A record is basically a "historical" document that contains information worth keeping for some time. The most familiar form of record keeping is the documentation we maintain for the IRS. And for an FDA-regulated organization, the need to maintain lot history records is made painfully clear via United States Government penalties.[5]

Records as Objective Evidence

However, the ISO 8402:1994 vocabulary defines a record as a

➪ "document which furnishes *objective evidence* of activities performed or results achieved."

This definition significantly expands the classification of documents as records and requires some discussion.

To Initiate the Records Master List

An excellent rule of thumb used to form the records master list is to first list all of the forms used by the organization. Invariably most of those forms will be kept by individuals in their files as a record of their acceptance, rejection, verification, identification, and categorization of their activities. Then, add in the more subtle records such as management reviews, design reviews, responses to requests for proposals and quotes, and preventive action reviews.

Specific Records

The Standard does call out a specific number of required records. However, the classification is generic. As a result, it is necessary to interpret the spirit of the requirement. The shock on the face of the records steward, when the assessor asks to see a list of documents contained in the contract review files, is not necessary.

Just remember that an assessor cannot assume what belongs in those files. The assessor must audit against what has been declared by the supplier as a quality record. The declarations can be challenged if either they do not comply with the Standard or it is discovered that many documents are kept as records but are not included in the master records list.

Table 4.1, entitled "ANSI/ISO/ASQC Q9001-1994 Record Implications," summarizes the explicitly called for records as well as the implied (implicitly) required records for several elements of ISO 9001. Appendix B is a summary of all the elements. In addition, Table 4.1 expands on the generic requirement to give the reader some idea of the pitfalls inherent in this clause.

Notice that "master lists" are included. Such lists are normally locally controlled documents that require a name and a date. However, they also contain information—for example, current revision level, the location of the document, and to whom it is signed out—which the assessor requires for objective evidence of, in this case, document and data control. This puts the lists into the category of a "record."

Records Quantity

The number of records maintained by an organization is always a "bone of contention" with top management. Unfortunately, the harder you push to remove records, the more you will find. People need them to do their jobs: Just make sure they are useful.

TABLE 4.1
ANSI/ISO/ASQC Q9001: 1994 Record Implications

ISO 9001 Element	Explicit Records	Implicit Records	Typical Records Required
4.1	4.16		▪ Management reviews ▪ Performance graphs (metrics) ▪ Customer satisfaction or dissatisfaction
4.3	4.16		▪ RFPs/RFQs ▪ Sales orders ▪ Customer purchase orders ▪ Contract reviews & approvals ▪ Invoices ▪ Customer complaints ▪ Customer specifications
4.4	4.16		▪ Assignment of personnel ▪ Project plans ▪ Project interface documents ▪ Customer design inputs ▪ Design specifications ▪ Design reviews ▪ Design verification results ▪ Design validation results ▪ Design changes
4.5		4.16	▪ Master document control lists ▪ Change notices ▪ Master engineering lists ▪ Retained obsolete documents ▪ Nature of the changes
4.6	4.16		▪ Approved subcontractor lists ▪ Subcontractor monitoring ▪ Purchase orders ▪ Quality requirements ▪ Source inspection records

Scrap Paper/Computer Transfer Notes Are Not Records

It is also important to know when a form is maintained for a significant reason, because many forms are really "scrap paper" and need not be controlled. A good example of this is in a basically on-line system in which raw data are fed into a computer data base at the end of each day. Obviously, the form used to collect the raw data could have been a notebook page and can be discarded. The "record" is controlled on electronic media.

PART 3: SPECIAL MANDATORY REQUIREMENTS

Customer Complaints as a Mandatory Requirement

In its charge to the accredited registrar, EN 45012, Par 18 states that

➪ "The Certification Body *shall* require the certificated suppliers to keep a record of all complaints and remedial actions relative to the Quality System."

The Standard states in 4.14.2 Corrective Action that

➪ "The procedures for corrective action *shall* include: a) the effective handling of customer complaints and . . . "

We have generally found this set of directives misunderstood. What is called for is a clear statement in regard to how customer complaints are managed and recorded. This could be done by means of corrective action reports, marketing and sales logs and memos, or any combination thereof. For those who work in the medical industry, this directive is simply part of the general FDA/CGMP requirement.

The discussion on customer complaints is usually found in the section that responds to element 4.14 Corrective and Preventive Action.

Factored Items

We can define a factored item as a product that is shipped to your customer with your logo on it but which was not manufactured under your certified quality management system.

Example of a Factored Item

An example of a factored item is a can of some chemical that you might stock and sell to your customers for their convenience. The can is purchased under private label from the manufacturer, is either inventoried in your shop and shipped from stock, or drop-shipped to the customer by the manufacturer. Although you do check the label for accuracy, you do not verify the product's specifications or integrity. The sale of such cans represents a significant percent of your total sales, for example, 1 percent or more of total revenue.

As a result, this product must be declared as a factored item and cannot be included under those products that are processed through your ISO 9001 quality management system. Your customers will need to be made aware of this a situation in some manner. In such cases, it is best to inform your registrar of the issue so that a decision can be made on the best course to follow. The registrar will need to make the final decision.[6]

The remedy could be just a simple sampling plan, or it could be as serious as a resident QC inspector at the manufacturer's plant. Other methods include periodic vendor audits and certificates of compliance or analysis.

The declaration of a factored item(s) is usually placed in the section which discusses the element 4.2 Quality System.

Currency of Standards and Codes

The Standard is somewhat nebulous in regard to required standards and codes. There are several statements that, taken as a whole, give some indication of the intent:

- ➪ Par 4.2.3 g) Quality planning—" . . . the clarification of Standards of acceptability for all features and requirements, including those which contain a subjective element; . . ."
- ➪ Par 4.4.4 Design input—"Design-input requirements relating to the product, including applicable statutory and regulatory requirements shall be identified. . . ."
- ➪ Par 4.9 c) Process Control—"Controlled conditions shall include the following: compliance with reference Standards/codes. . . ."

What is advisable and appears to be generally acceptable to registrars is to create a list of the product/process-oriented standards and codes maintained by the organization, state who is responsible for them, show where they are kept, and explain how they are kept current.

Table 4.2 entitled "Excellent's List of Current Standards and Codes," summarizes what a typical list might look like. Notice that only standards that affect process or product are listed; for example, OSHA is not.

PART 4: SUMMARY OF GLOBAL MANDATORY REQUIREMENTS

This ends our review of the global "mandatory" documentation requirements for ANSI/ISO/ASQC Q9001-1994. We have shown that the following global documentation is mandatory:

- A quality system
- A quality manual
- Procedures (Tier II documentation)
- Method of linkage between Tiers
- Declaration of the ISO management representative
- Declaration of the organization's quality objectives
- Description of the organization to be certified
- Discussion of responsibility and authority of at least the top management
- "Inter-" and "intra-" organizational interfaces
- Demonstration of the effective implementation of the system
- The use of sensible levels of documentation
- Records to indicate objective evidence of effective operation
- The effective management of customer complaints
- Declaration of factored items (if applicable)
- Master list of current Standards and codes

For completeness we will now cover the optional requirements.

Table 4.2
Excellent's List of Current Standards and Codes

Name of Standard and Code	Responsible Manager	Where Located	Currency Method
21 CFR, Part 210	Regulatory affairs	Office files	U.S. Government publications
ANSI/ISO/ASQC Q9001-1994	ISO 9000 management representative	Office files	Subscription to the ASQC
AAALAC guidelines	Pathology	Laboratory files	AAALAC committee member
NIH guidelines	Operations	On-site files	Subscription
U.S. Pharmacopeia National Formulary	Quality assurance manager	Laboratory files	Subscription to USP-NF
IPC-610 manufacturing codes	Quality assurance manager	Design and manufacturing areas	IPC membership
CE Mark – EN60601-1-2 and EN55011 Class B IEC 801-2 IEC 801-3 IEC 801-4 IEC 801-5	Engineering design administrator	Administrator's office files	Review annually obtain latest revisions from test house.
UL 2601	Design engineering	Design engineering files	Subscription
Big Three automotive standards	Design engineering	Design engineering library	Supplied by Big Three
Mechanical contractor standards	Senior engineer	Contractor's engineering library	Subscription

Endnotes

1. The exponential increase of electronic media based systems is concurrent with the explosion in information technology. As a result, there are numerous articles published each month on the subject. Here are two examples: "Computer-based Systems Streamline Device Manufacturing," Michael J. Major, *Medical*

Device & Diagnostics Industry, November 1996, p. 67; and "Managing Design Data: The Five Dimensions of CAD Frameworks, Configuration Management, and Product Data Management," Peter Van Den Hamer, & Kees Lepoeter, *Proceedings of the IEEE,* Vol. 84, No.1, January 1996, p. 40.

2. See for example, C. Michael Taylor, "Bored with the Same Old Standards Books? *Automotive Excellence,* Spring 1997, p. 6, A Publication of the ASQ Automotive Division.
3. As with all powerful ideas, an ISO 9000 mythology has already been created in spite of its short, ten-year existence. One of the myths is that each Tier requires a document. We feel that not only is such an approach contrary to the spirit of the Standard, but the redundancy that results from such a viewpoint is counterproductive and serves to confuse the users instead of support their efforts.
4. The subject has begun to get some deserved dedicated attention. See, for example, "Managing Records for ISO 9000 Compliance," Eugenia K. Brumm, *Quality Progress,* January 1995, p. 73. The hardcover book of the same title is also available from ASQ, ISBN 0-87389-312-3.
5. Quality System Regulation, Part VII, Department of Health and Human Services, Food and Drug Administration, 21 CFR Part 820, Current Good Manufacturing Practices (CGMP): Final Rule, October 7, 1996, published by the SAM Group, Stat-A-Matrix, Edison, NJ, Sec. 820.181 Device Master Record.
6. Although the supplier is ultimately responsible for the choice of contractual standard and how that standard is applied in the organization, it is essential to keep in close touch with your registrar on interpretation because registrars have the same issue—how should they interpret the Standard against the requirements of their accreditation board.

CHAPTER 5
OPTIONAL REQUIREMENTS

PART 1: ISO 9000 QUALITY PLANS—OPTIONAL

The optional requirement for quality plans is stated in 2.2.3 a) , Quality Planning:

- "The supplier shall give consideration to the following activities, as appropriate, in meeting the specified requirements for products, projects or contracts: a) the preparation of quality plans;"

A quality plan is defined in ISO 8402:1994 as

- "A document setting out the specific quality practices, resources and sequence of activities relevant to a particular product, service, contract, or project."

About forty years ago, quality plans were very common in MIL-Q-9858 quality control systems and consisted of "bubble" flow charts with all of the associated documentation affixed to the chart. Today, quality plans vary greatly and are an integral part of the QS-9000 requirements,[1] and are discussed in some detail in ISO 10005:1995(E).[2]

Looks Like a Duck

In light of the large number of books published on the "mind," we do have to slightly modify the old adage, "If it looks like a duck, walks like a duck, and sounds like a duck," it could be either a real duck or a Turing machine.[3] At any rate, a "quality plan" sure sounds like a process, and indeed it is, i.e., the vocabulary for ISO Standards defines *"process"* in the following way:

ISO 8402:1994(E/F/R):

- "set of inter-related resources and activities which transform inputs into outputs
- Note—Resources may include personnel, finance, facilities, equipment, techniques and methods."

As a result, the old "bubble" chart configuration is as true today as it was forty years ago and is a very useful rule of thumb in the creation of a quality plan graphic (refer to Figure 5.1, entitled "Process Flow with Documentation = Quality Plan.").

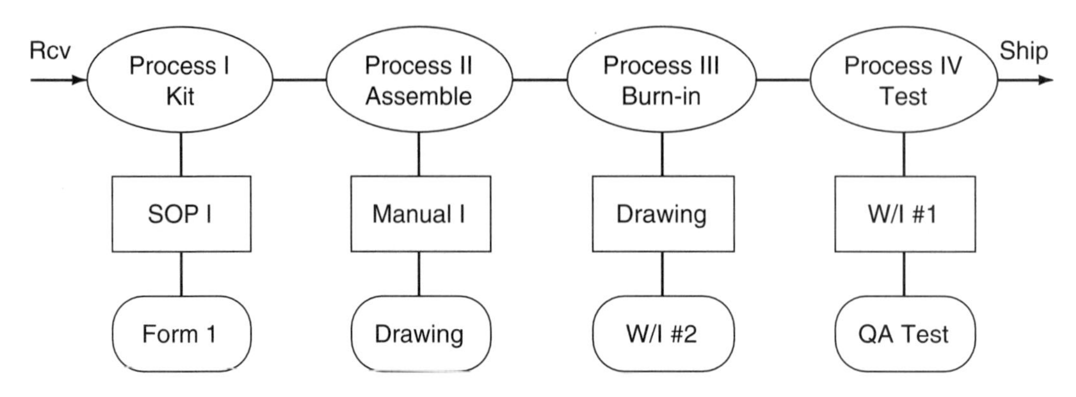

Figure 5.1 Process flow with documentation = quality plan.

However, we have termed the quality Plan as "Optional" for several reasons. First, the Standard uses the term "as appropriate," and second, the authors knew very well that there were few of us left who remember the old charts and therefore added as Note 8 a safety-valve:

➪ "The quality plans referred to (see 2.2.3a) may be in the form of a reference to the appropriate documented procedures that form an integral part of the supplier's quality system."

Accordingly, it is clearly permissible to define the quality plan policy statement with "We use the documented and referenced procedural system in lieu of quality plans," if quality plans are not a contractual requirement.

PART 2: WORK INSTRUCTIONS—OPTIONAL

Another option appears as an note at the end of paragraph 2.2.2 of the Standard:

➪ "Tier III Note—Documented procedures may make reference to work instructions that define how an activity is performed."

This was a mandatory condition in the 1987 version of the Standard and was lowered to a note because of its lack of universality. However, work instructions are often found in organizations in the form of, for example, process sheets, test procedures, and purchasing procedures. Work instructions can also be found embedded within a high-level procedure. A possible example of this format is shown in Table 5.1, entitled "Integrated Process and Work Instruction Document."

In this table we show an abbreviated three-phase front-end process for a company that receives a certificate of analysis with some of its raw material, for example, steel rods. The process flow is from receiving to inventory to kitting. Each part of the process flow has several activities. The several activities have a number of work instructions shown in the far-right column. The forms required are also noted, as well specific responsibility for the activity.

An integrated format of this type is very powerful in terms of simplicity and necessary detail.

TABLE 5.1

Integrated Process and Work Instruction Document

Process Phase	Activity Description	Work Instructions
1.0 Receiving	1.1 Receiving—at dock (receiver) 1.2 Receiving inspection (QA)	1.1.1 Log in the material (F-1.1.1) 1.1.2 Check for the *C of A* 1.1.3 Send *C of A* to QA 1.2.1 Compare material properties to *C of A* and log-in (F-1.2.1) 1.2.2 Use RTV form if not acceptable (F-RTV-1)
2.0 Inventory	2.1 Inventory—production control 2.2 Materials—movement coordinator	2.1.1 Enter material into the MRP system—note *C of A* received 2.1.2 Release modified schedules to production 2.2.1 Move *C of A* material to special storage 2.2.2 Log-in (F-2.2.2)
3.0 Kitting	3.1 Kitting—materials control 3.2 Kit-release coordinator	3.1.1 Check to see that *C of A* was received 3.1.2 Obtain pick-list 3.1.3 Compare to BOM 3.2.1 Final inspection of kit (F-3.2.1) 3.2.2 Attach traveler (F-3.2.2) 3.2.3 Release kit to assembly

Endnotes

1. See for example, *Advanced Product Quality Planning (APQP),* Reference Manual, from AIAG, June 1994, Ph:810-358-3003.
2. ISO 10005:1995(E) presents several typical quality plan configurations that include plans for service organizations, manufactured product, processed material, and a software life cycle.
3. After Alan Turing, the sometimes recognized inventor of the digital computer. See for example, *The Emperor's New Mind,* Roger Penrose, Oxford University Press, NY, 1990.

Chapter 6
"Shall" Analysis

Part 1: Expressions of Policy

Based on ISO 8402:1994, Quality management and quality assurance—Vocabulary, the organization's quality policies (operational rules of the house) are to be stated in the Manual.[1]

ISO 8402:1994 defines such policies as the

➪ "overall intentions and direction of an organization with regard to quality, as formally expressed by top management."

We have found that the most intense areas of interpretive conflict center about

- What needs to be expressed as policy (policy scope)
- The level of detail expressed within the policy statements
- The use of the title, "Quality Policy Manual," when the Quality Manual is a stand-alone document that does not contain the lower-tier documents.

The first two issues require resolution. The third is a matter of choice, and we leave it at that because the use of either title is incidental to the integrity of the QMS. We will first discuss the scope of the policy statements.

Scope

It is common to find ISO 9000 manuals written in a sequenced form that corresponds directly to the Standard's twenty elements, for example, a manual with twenty sections, each section entitled the same as the Standard's title, as

- Section One—Management Responsibility
- Section Two—Quality System
- Section Three—Contract Review . . .
- . . . Section Twenty—Statistical Techniques

Our thesis requires that each of the twenty sections should respond in detail to each "shall" within the corresponding element. We now wish to determine just how many "shalls" there are so that we can estimate the scope of our effort.

The number of effective "shalls" per element is summarized in Table 6.1, entitled, "Effective "shalls" per ISO 9001; 1994 Element." As indicated, although there are 138 explicitly stated "shalls," there are actually 320 expanded "shalls" that require a response.

The expansiveness versus the Standard's explicitness is one of the more subtle and difficult parts of manual creation process. One often wonders, "Where in the world did the assessor come up with that question?" An experienced assessor simply goes through all 320 expanded "shalls" specified in the ANSI/ISO/ASQC Q9001-1994 International Standard, one at a time, consciously or unconsciously. As you can see, this would involve what seems to be an endless progression of interview questions.

By the fourth or fifth surveillance the depth of questions even surprises the assessor who produces them.

Of course, the exact number of expanded "shalls" is not the point and the exact number is subject to all-night debates. It is the awareness of the completeness of the Standard that is significant. Those who created the Standard really did a fine job of it.

We believe, that the Standard, when fully responded to, establishes a powerful platform for total quality management.

As we have indicated in Table 6.1, this expansive nature of the Standard occurs in numerous elements. The use of this table—and its expanded version in Appendix C—becomes a useful guide when internal audit checklists are created.

Method to Count "Shalls"

We will now analyze the first sentence of 4.1.1 Quality Policy of the Standard to demonstrate how we have counted this expansive statement of "shalls."

- "The supplier's management with executive responsibility shall define and document its policy for quality, including objectives for quality and its commitment to quality."

4.1.1 Analysis—Notice that there is only one explicit "shall" in the clause. However, also notice that the clause expands into three directives that require a response, not just one, that is

- shall define and document its *policy* for quality;
- shall define and document its *objectives* for quality; and
- shall define and document its *commitment* to quality.

To illustrate this technique further, we look at clause 4.2.1 Quality System—General.

- "The supplier shall establish, document, and maintain a quality system as a means of ensuring that product conforms to specified requirements.
- The supplier shall prepare a quality manual covering the requirements of this international Standard.
- The quality manual shall include or make reference to the quality system procedures and outline the structure of the documentation used in the quality system.
- Note 6 'Guidance on quality manuals' is given in ISO 10013."

Table 6.1
Effective "Shalls" per ISO 9001:1994 Element

ISO 9001 Clause	Explicit Shalls	Shalls by Expansion
4.1	9	42
4.2	9	23
4.3	5	11
4.4	19	33
4.5	7	14
4.6	10	20
4.7	2	7
4.8	3	5
4.9	8	14
4.10	15	21
4.11	9	30
4.12	2	2
4.13	7	14
4.14	5	17
4.15	9	13
4.16	7	18
4.17	6	15
4.18	3	6
4.19	1	5
4.20	2	10
Totals	**138**	**320**

4.2.1 Analysis—In this case we have three explicit "shalls." However, when we expand the "shalls" we get six:

- ➪ The supplier shall *establish* . . .
- ➪ The supplier shall *document* . . .
- ➪ The supplier shall *maintain* . . .
- ➪ The supplier shall *prepare* a quality manual . . .
- ➪ The quality manual shall *include or make reference* . . .
- ➪ The quality manual shall *outline the structure* . . .

It is of interest to note that this simple exercise, in practice, permits the reader to grasp for the first time the nuances of the Standard. What at first appears to be a document in which every sentence looks the same and is undifferentiated from any other sentence, suddenly takes on the appearance of a good mystery novel as we start to look for the plot. We assume, of course, that you enjoy mystery novels.

One thing is clear: It helps if you happen to be either a Talmudic or Biblical scholar.

Part 2: Section Length

There is some correlation between the number of expanded "shalls" and the number of pages in a given manual section, but it is a weak one. The actual number of words required in a section has more to do with the scope of the "shall," although Section 4.1 Management Responsibility is generally the longest, followed by 4.4 Design Control and/or 4.2 Quality System.

A good example of the inherent lack of preciseness in such an estimate is to consider that for a service organization whose specialty is repair and calibration of test equipment, Section 11—Control of Inspection, Measuring, and Test Equipment can represent over 40 percent of a manual.

Elements 4.1 and 4.2 require some graphics in the form of charts and figures, which also increase the number of pages in those sections, although appendices can be used to decrease the explicit size of each. A graphic for elements 4.4 Design Control and 4.9 Process Control is common if the system is online.

Finally, font size differences make a study of such relationships almost meaningless.

- The rule of thumb is—say what is necessary, regardless of length, and use a font large enough to be read with ease.

The smallest, fully responsive Manual for an accredited and certified company we have ever seen was approximately 4 × 6 inches square, 33 pages long, and typed in a font size of no more than 8 point. It was a little difficult to read, but every employee knew about the Manual and what was in it.

Endnotes

1. The ISO Vocabulary speaks of *policy* in singular terms, that is, " . . . document stating the *quality policy* . . . ," and in Par "3.1 Quality Policy." It is necessary to refer to the ISO 10013 guideline to realize that the entire twenty elements of the Standard define the quality policy, for example, in Par "4.2.2 Structure and format. . . . One of the methods of assuring that the subject matter is adequately addressed and located would be to key the sections of the quality manual to the quality elements of the governing quality system Standard. . . ." Thus, the quality policy consists of at least twenty policy statements and specific responses to the requirements (shalls) called for in each element.

CHAPTER 7
CONCOMITANCE

PART 1: ASSOCIATED ELEMENTS

Each "shall" denotes a specific requirement of the Standard. The requirements are linked in such a way that they are often associated conditions for one another. As a result, if any one "shall" is not adequately addressed there is an impact on other sections of the Standard because each "shall" is a part of the Standard's overall fabric.

In addition, each ISO 9001 "shall" is written to be "descriptive," with the intent that we are to reply in a "prescriptive" manner:

- By *"prescriptive"* is meant that the responses include operational details.

In that way the Manual can be used effectively by decision makers, for example, customers who must decide on whether or not to audit their subsupplier.

Training

An example of this concomitant—or interelement—relationship concerns "training" that occurs explicitly in elements 4.1 Management Responsibility, 4.2 Quality System, 4.9 Process Control, and 4.18 Training of the Standard. Table 7.1, entitled "Associated Training Relationships in ANSI/ISO/ASQC Q9001-1994," summarizes this set of conditions.

In each case shown in Table 7.1, "training" is viewed from four different perspectives, yet they are synergistic, that is

- ➪ in 4.1 we discuss training as a top-down budgetary issue.
- ➪ in 4.2 we consider the impact of training on the technical depth of our documentation.
- ➪ in 4.9, personnel who work with "special" processes, for example, those that produce product that cannot be fully verified and validated before shipment, are to be qualified for such work via training and
- ➪ in 4.18 we deal with the entire training program and its operational imperatives.

A more general perspective can be gained if we look at the total set of twenty elements at a glance as summarized in Table 7.2, entitled "ANSI/ISO/ASQC Q9001-1994 Concomitance at a glance."

Table 7.1

Concomitant Training Relationships in ANSI/ISO/ASQC Q9001:1994

ANSI/ISO/ASQC Q9001-1994 Clauses	In That Clause We Are Required to
4.1.2.2 Resources	. . . "identify resource requirements and provide adequate resources, including the assignment of trained personnel, for management, performance of work, and verification activities including internal quality audits."
4.2.2 Quality System Procedures	" . . . manage the range and detail of our procedures such that they depend upon the complexity of the work, the methods used, and the skills and training needed by personnel involved in carrying out the activity."
4.9 Process Control	" . . . The requirements for any qualification of process operations, including associated equipment and personnel (see 4.18), shall be specified . . . "
4.18 Training	" . . . identify "training needs and provide for the training of all personnel performing activities affecting quality." In addition, "Personnel performing specific assigned tasks shall be qualified on the basis of appropriate education, training and/or experience, as required. Appropriate records of training shall be maintained."

In this table we indicate the cases where an element of the Standard clearly references another element, for example

- when clause 4.1.3 Management Review refers directly to 4.16 Control of Quality Records; and
- where the reference is implied, for example, clause 4.2.3 Quality Planning specifies that
  - ➪ "The supplier shall define and document how the requirements for quality will be met," which implies that a statement is required as to just how the methods of resource planning (clause 4.1.2.2 Resources) cover the quality planning requirements.

The table also includes specifically referenced ISO 9000 guidelines distributed as notes throughout the Standard.

Another of the more subtle implicitly stated associations is found in clause 4.19 Servicing where it is stated

- ➪ "and reporting that the servicing meets the specified requirements."

TABLE 7.2

ANSI/ISO/ASQC Q9001: 1994 Concomitance at a Glance

ISO 9001 Element	Explicit Concomitance	ISO 9000 Noted Guideline(s)	Implicit Concomitance
4.1 Management Responsibility	4.16, 4.18		
4.2 Quality System	4.16	ISO 10013:1995	4.1.2.2, 4.4.2, 4.9 f), 4.10
4.3 Contract Review	4.16		4.4.4
4.4 Design Control	4.16, 4.3		
4.5 Document & Data Control			4.2, 4.4, 4.16
4.6 Purchasing	4.16		
4.7 Control of Customer-Supplied Product	4.16		
4.8 Product Identification & Traceability	4.16		
4.9 Process Control	4.16, 4.18		4.2 g)
4.10 Inspection & Testing	4.16, 4.13		
4.11 Control of Inspection, Measuring & Test Equipment	4.16	ISO 10012-1:1992	
4.12 Inspection & Test Status	4.13.2		4.16
4.13 Control of Nonconforming Product	4.16		
4.14 Corrective & Preventive Action	4.16, 4.1.3		
4.15 Handling, Storage, Packaging, Preservation, & Delivery			4.16
4.16 Control of Quality Records			4.6, 4.16
4.17 Internal Quality Audits	4.16	ISO 10011-1:1990 ISO 10011-2:1991 ISO 10011-3:1991	
4.18 Training	4.16		4.1.2.2, 4.2.2
4.19 Servicing			4.16
4.20 Statistical Techniques			4.16

which implies that if reports are required they will need to be kept as records to prove that the report protocol was actually followed (objective evidence). This type of implicit requirement is a common thread throughout the ISO 9001 Standard and is one of the reasons that interpretations do vary considerably among practitioners.

Total Response

- As a result, to ensure total system concomitance, we maintain that it is necessary to respond to every "shall" in a given ANSI/ISO/ASQC Q9001-1994 element and to present this information in the Quality Manual.

PART 2: NONAPPLICABILITY

In those cases where a "shall" is not applicable—that is, not appropriate to the organization's structure—this necessitates a discussion that clarifies the reason for its dismissal.

Positive Approaches to Conflicts

Clause 4.15.6—Often, a positive approach can be used to clarify the nonapplicability of a clause and to state what actually occurs operationally.

Consider for example, a response to 4.15.6 Delivery of the Standard, in regard to a contractual agreement to extend product quality protection beyond F.O.B. the plant, where for some reason this cannot be legally or economically performed.

A positive approach would be to state that

- **Positive**—"The Excellent Corporation considers product quality paramount in our customer relationships and the level of protection of all product shipped is determined by quality assurance laboratory testing to the appropriate international Standard, as necessary."

This type of statement puts the customer at ease and implies that you are someone they can negotiate with productively.

Contrast this approach with the alternative dogmatic statement

- **Negative**—"F.O.B. is always established at the Excellent Corporation's dock."

Now customers will wonder what difficult type of negotiations they are apt to run into with this supplier who sees the world without shades of gray.

Clause 4.6.4.1—The same approach can be used in the case of clause 4.6.4.1 Supplier Verification at Subcontractor's Premises of the Standard. The positive response might be

- **Positive**—"The Excellent Corporation does not consider it necessary to perform source inspection on our vendor's products since they provide certificates of compliance with each shipment."

As opposed to

- **Negative**— "The Excellent Corporation does not require source inspection."

The customer will wonder just how you really control your subcontractors—or do you?

CHAPTER 8
APPROPRIATE DETAIL

PART 1: DETAIL LEVEL

There is also considerable debate in regard to the detail required in an ANSI/ISO/ASQC Q9001-1994 policy statement. We have found that the conflict lies in the belief that the level of detail is an appropriate way to measure policy effectiveness.

We wish to demonstrate that the level of detail varies widely with policy statements, and that it is of prime importance that enough detail be available so that the readers can use described rules and methods to intelligently make business decisions that impact their organizations.

CASE IN POINT—AN ISO 9000 CERTIFIED VENDOR

Every day, buyers and quality personnel make joint decisions on whether or not to add a new key vendor to their approved supplier list (ASL). In those cases where the new supplier is ISO 9000 certified, one of the decisions that can be made is to avoid the expense of a vendor audit and rely on the depth and scope of the supplier's Quality Manual for the decision on whether or not to add the vendor to the (ASL).

We have been there when the Quality Manual arrives and is so nebulous that it cannot be used as part of the decision process. The first response is to laugh about it and wonder how in the world the vendor was ever certified. The final response is anger, because now thousands of dollars and several precious days must be spent to run the vendor audit.

The final blow comes when the audit is run. The vendor turns out to have a perfectly fine shop—another case in which the Manual does not reflect the competency of the supplier. A very common event.

Thesis

We will demonstrate that the level of detail in a policy statement is an inappropriate way to decide on the effectiveness of the statement.

EXAMPLE #1—ON STANDARDS AND CODES

First Statement—The following is a broadly stated policy statement:

- "All The Excellent Corporation employees shall comply with reference Standards and codes."

Second Statement—This is also a policy statement on the same subject, but it is definitive:

- "All Excellent Corporation employees wear safety glasses and ear plugs before entering designated manufacturing areas."

Example #1 Analysis—The first statement is a philosophical directive equivalent to reading back (or paraphrasing) the International Standard. It uses the future tense, so it is not clear that the rule is actually in place yet. It represents a common paraphrasing technique.

The second statement has sufficient detail to be clear for anyone reading it as to what the organization expects—and it can be readily audited. This statement is presented in a way that is similar to the phraseology used in ISO 10013:1995 Quality Manual Guideline, page 11.

EXAMPLE #2—ON TEST MEASUREMENTS

First Statement—A paraphrased policy statement:

- "The Excellent Corporation shall determine the test measurements to be made and the accuracy required."

Second Statement—A definitive policy statement:

- "The Quality Assurance Department of Excellent Corporation establishes the degree of acceptance testing required on all products, and documents such tests on specification work orders.
- All inspection, measuring, and test equipment purchased for this purpose is reviewed and approved by the quality assurance department for the accuracy and precision required. A 4:1 rule is generally used, if applicable, to determine instrument accuracy."

continued

Example #2 Analysis—The first statement yields neither responsibility nor method. The second statement is very clear and provides the reader with the type of knowledge necessary to either choose the vendor or prepare a productive audit of the vendor. One gets the feeling of competence and completeness.

Example #3—On Internal Quality Audits

First Statement—A broad policy statement improved somewhat over a paraphrased one:

- "At Excellent Corporation, internal quality audits shall be scheduled based on how important the activity is to the company and how well it has performed in previous audits."

Second Statement—A definitive policy statement prescribed in ISO 10013:1995(E), page 11, Par 4.17.4.2.

- ➪ "The scope of the audits is determined with regard to the importance of the activities in question and the knowledge of any existing or likely problems."
- ➪ "The audit frequency is, at least: for quality system audits—once per year; for product quality audits—twice per year; for process quality audits—once per year. Audit plans are made up and documented once per year. Checklists are prepared as an aid."

Third Statement—A definitive policy statement that specifies responsibility and avoids specific numbers—which could vary widely during the year:

- "At Excellent Corporation, all of the elements of the Standard are audited against every department, as applicable, on a yearly basis. The director of manufacturing establishes the audit plan and issues the internal quality audit schedules—which include vendor audits—as required."
- "The frequency with which a given department is audited is based on the results of previous audits. The total audit program includes systems, process, and product audits. All audits require checklists prepared prior to the audit by the auditors and approved by the lead auditor."

Example #3 Analysis—The first statement is philosophical. The second and third statements are detailed, yet concise, and loaded with information about the company. The reader has little trouble in visualizing the depth and scope of the audit program in either Example #2 or #3.

PART 2: IN PRACTICE

In our review of over seventy Manuals, we have found a tendency to avoid detail in the policy statement based on the belief that the Manual should be of a certain size, and to essentially "throw out the baby with the bath water."[1]

The inappropriateness of this belief in a specific length (size) is self evident when you consider that ISO 9000 Manuals are written for organizations with from two to ten thousand employees. In a very large company, the organization chart appendix can be half as long as the Manual for a company with eight employees.

Often, the policies end up in the next-tier document, so there is an awareness that the information is required, but there is also an issue of where it belongs. Sometimes, the quality policy statements appear nowhere, much to the chagrin of the ISO 9000 management representative who searches with great gusto and futility to prove the existence of a response that was believed to be too complex to place in the Manual.

Conclusion

In summary, we maintain that a quality policy statement should be

- a direct response to every "shall" in the Standard
- stated in present tense as opposed to future tense
- clearly expressed in simple declarative prose
- not paraphrased
- of whatever length and detail is necessary to define the organization's rules and methods

We conclude that the level of detail in a policy statement should be whatever the "shalls" demand in context with the methods used by the organization. Short policy statements are not necessarily effective and are often inappropriate.

Endnotes

1. The controversy on Quality Manual design appears to have reached a level of concern sufficient to cause publications on the subject, e.g., "How to Avoid Creating the Dreaded Big Honkin Binder," Richard Balano, *Quality Progress,* March 1997, p. 152.

SECTION 3

QMS STRUCTURE

"(George) Miller (1956) showed that the individual's ability to make absolute distinctions among stimuli, to distinguish phonemes from one another, to estimate numbers accurately, and to remember a number of discrete items all seemed to undergo a crucial change at about the level of seven items. Below that number, individuals could readily handle such tasks: above it, individuals were likely to fail. Nor did this discontinuity seem accidental."

The Mind's New Science, *by Howard Gardner*
Basic Books, Inc., Publishers, New York,1985, p. 89.

Chapter 9
The Documentation Pyramid

Part 1: Position of the Quality Manual in the Pyramid

Introduction

In Section 2, we illustrated the mandatory QMS requirements and demonstrated the way to effectively transpose those requirements into quality policy statements. We now wish to present the mechanics of an effective QMS. However, as we were warned by George Miller in 1956—and which has since proven to be a most powerful observation—it is important that we constantly remain alert and keep the portions of documentation as small as possible. Because of the 20 elements (portions) in ISO 9001, this is not always accessible to us. As a result, we require an increased level of training beyond which we would normally consider.

We will assume that the reader is committed to this relatively high level of employee training and has mastered this art. We begin our discussion with the flow of information throughout the QMS.

The Manual

The Quality Manual (Manual) stands as a colossus above the ISO 9000 documentation hierarchy and sources the flow of QMS information – refer to Figure 9.1, entitled "The Four-Tier Operational Pyramid Concept-ISO 9000 Guidelines." Please note that, for the sake of clarity, we have chosen to discuss the stand-alone quality policy manual configuration because it is the most common one used. The alternative configurations are discussed in Part 3 of Chapter 11, entitled "Manual and Configurations."

Guidelines

As we have observed in the previous section, the tiers and various examples of documents are merely *guidelines.* It is only the Quality Manual, Procedures, and Records that were mandatory hierarchal documents in ANSI/ISO/ASQC Q9001-1994.

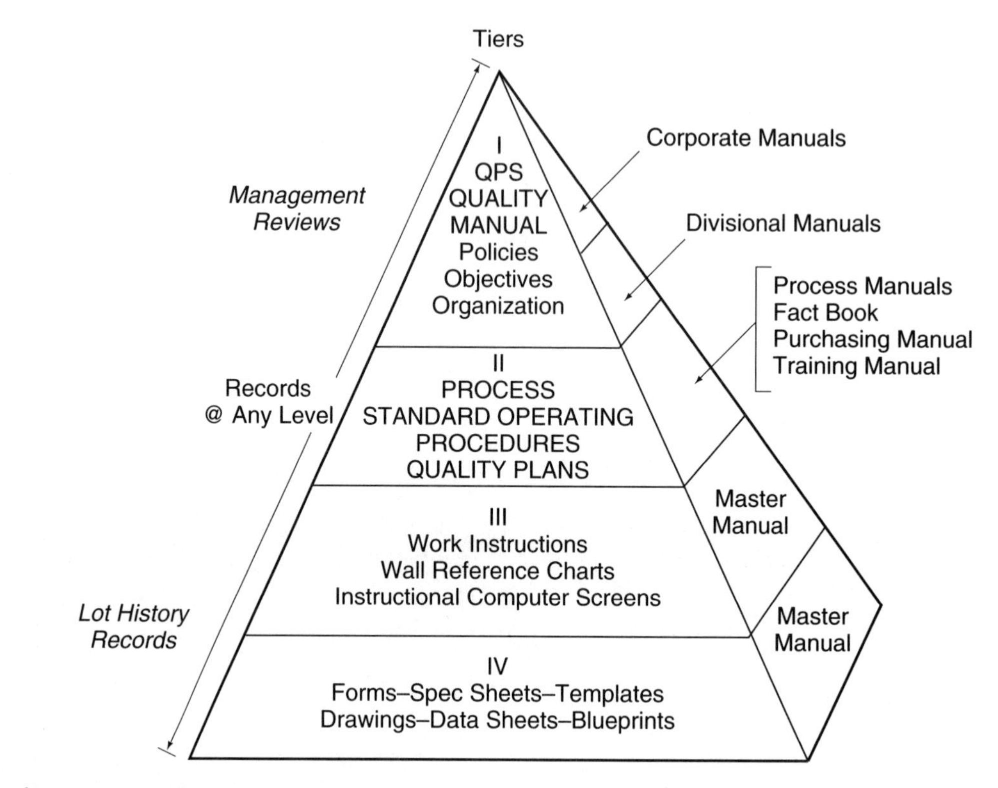

Figure 9.1 The four-tier operational pyramid concept—ISO 9000 guidelines.

Four Tiers

We have chosen a four-tier operational pyramid to illustrate the impact of the Manual on the entire documentation structure, although—as we note—the pyramid is meant to be only a guideline since it does not replace the actual linkage that must be present from document to supplemental document.

The four-tier structure was chosen because it readily provides levels for the type of documents that we have usually encountered. However, some companies have as few as two defined levels and some as high as six defined levels. The number of levels is irrelevant. What is relevant is that they are presented in a way that aids the reader to navigate easily throughout the system.

Part 2: User-Friendly

The Four-Tier Pyramid illustrates how important it is to have a Manual that does provide user-friendly navigation. This direction is given in clause 4.2.1 General of the Standard:

➪ "The supplier shall prepare a quality manual covering the requirements of this International Standard.

➪ The quality manual shall include or make reference to the quality-system procedures and outline the structure of the documentation used in the quality system."

As a result, the Manual should tell the reader at least the following structural information:

- The number of tiers chosen, their contents, and the method of labeling
- The method of linkage, for example, by reference numbers in the text, a document tree
- Whether the system is hard copy, on electronic media, or a mixture
- The type of documents to be found, for example, in manuals, on-line documentation, individual copies, wall reference charts
- How quality management documents, for example, the Manual, are differentiated from engineering documents, for example, drawings

Matrix Format

The four tiers can also be described in the form of a table (see Table 9.1, entitled "The Four Suggested Operational Tiers of ISO 9000 Documentation Guidelines"). The matrix form provides another class of information related to the specific content of a given Tier.[1]

For example, we note that the text in the Manual tends to be *time-independent*. These are statements of policy—rules—that do not imply movement or process. In contrast, the text in the lower-level documents is made up of *time-dependent* statements. The text in such documents generally implies process or procedure or movement, for example, from department to department, division to division, operator to operator, or operator step to operator step.

In the far-right column of Table 9.1 we also see comments that tell us how to deal with document content. As a result, if policy is presented in the Manual it need not be restated in the procedures.

Operational Tiers

We have specified in Table 9.1 that the documentation pyramid represents the operational flow of information, that is, day-to-day processes carried out by use of dynamic and current documentation. That is why "records" are not included in Tier IV. This is contrary to common usage, which we believe is incorrect from a hierarchal standpoint. We realize that this is a fine point, yet it causes considerable confusion among Manual designers.

Analytical Linkage

The point we wish to make is that because "records" are basically historical documents, they are not part of the operational linkage. They are certainly part of the analytical linkage, which forms an information context in parallel with the operational process.

The difference between operational and analytical linkage can be demonstrated if one considers that, during the build cycle, the operator refers to a work

TABLE 9.1

The Four Suggested Operational Tiers of ISO 9000 Documentation Guidelines

Tier	ISO 9000 Category	Content Description	Deals with
I	**Quality Policy Manual** Quality Policy Statement Management review records	▪ A time-independent document describing the organization's policies written in conformance with the chosen ISO 9000 Standard model.	▪ The organization's response to each "shall" ▪ The "rules of the house," the methods used to ensure compliance ▪ Defined responsibility
II	**Process Documents & High-Level Procedures** SOPs DOPs Business plans	▪ Time-dependent documents that describe either the overall processes of the organization or a combination of process and high-level procedures, e.g., contract review, design control, corrective & preventive actions.	▪ Purpose? What? When? Where? Who? and How? at a high level ▪ Flow of information from area to area, department to department, building to building
III	**Lower-Level Procedural Documents;** Wall ref charts, instructional computer screens, test procedures, purchasing procedures	▪ Time-dependent, detailed step-by-step work instructions on how to complete a task, e.g., at the operator or bench level ▪ Often integrated into Tier II documents	▪ How do you do the job? ▪ Tells the reader in a step-by-step fashion ▪ Provides the necessary data to perform the tasks
IV	**Unfilled in Forms or Formats,** Templates, blueprints, schematics, data sheets, specifications, drawings	▪ Documents that specify the data requirements called out in the various documents and/or specific data sources ▪ Many of the forms are used as records once they are filled in, although specific records are required at all levels ▪ Complementary documents to support work instructions	▪ The forms we use to demonstrate that a procedure requiring either data-taking or data-input was done ▪ The templates required to measure and fabricate

FORMAT—Tier IV
Operational Document

Format #10-03-49-03 P2

Format #10-03-49-03 P1
Name......................
Station.....................
Tempt.......................
Date..........................
Comments...............
....
....
....

Form

* Control the format if you care about anyone changing it on their own.
* Document control nos. are common.
* Used by an operator on a day-to-day basis.
* Linked by references in procedures.

FORMAT w/DATA
Record—Analytical Document

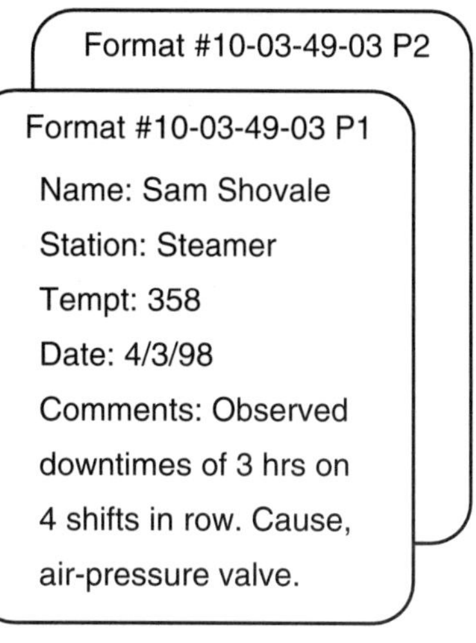

Record

* Control the completed format within the product, process, or project/program files.
* Another document control no. is unnecessary.
* Used by quality assurance for trend analysis.
* Linked via Quality Policy Manual to a high-level procedure from a master records list.

Figure 9.2 Operational versus analytical linkage.

instruction—a dynamic document. The operator does not refer to all of the travelers signed off over the past month to find out what to do in the next step.

However, the quality assurance manager analyzes all of the travelers over the past month to look for trends in nonconformances and time-related issues. The operator fills in the traveler form—a Tier IV document—and the quality assurance manager creates a nonconformance Pareto chart from the traveler records.

Forms versus Records

We have demonstrated this difference between forms and records in Figure 9.2, entitled "Operational versus Analytical Linkage."

By-Passes

It is unnecessary to create documents so that every Tier is covered; do what makes sense. In other words, if it is required to go from the Manual to a form, for example, management review format, do it. It is not necessary to create an SOP and a work instruction to get to that form. Figure 9.3, entitled "Possible Linkage Schemes," illustrates this concept.

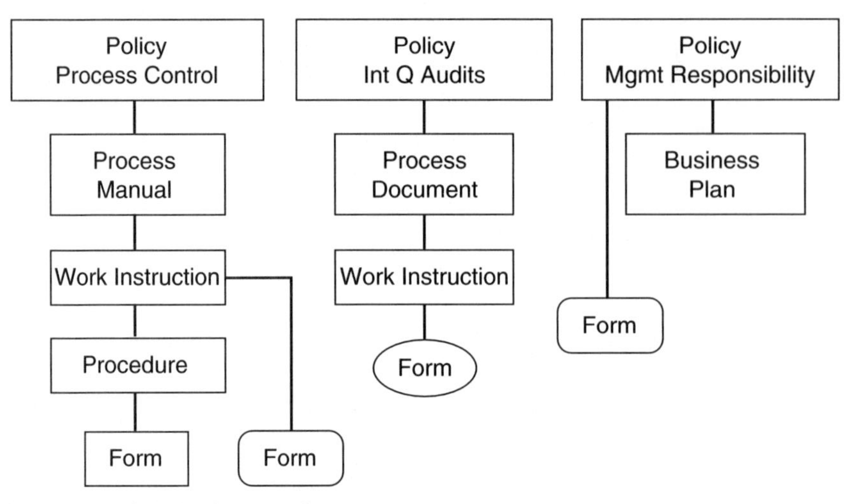

Figure 9.3 Possible linkage schemes.

PART 3: PYRAMID FOR A MANUAL

In a similar fashion we can describe the hierarchal content of the Manual as illustrated in Figure 9.4, entitled "The Four Phases of the Quality Policy Manual."

We indicate in Figure 9.4 that the Manual contains the entire set of organizational quality policies (defined as Phase 1). We have chosen to indicate twenty directly sequenced sections to cover the twenty elements of ANSI/ISO/ASQC Q9001-1994 (defined as Phase 2). However, as we will show in this section, Chapters 10 and 11, the same definitions are valid for any form of Manual sequences or configurations.

Phases 3 and 4 are somewhat more difficult to define since they are parallel processes in which each "shall" of the Standard (Phase 3) is responded to with a Quality Policy Statement (Phase 4).

- It is this four-phase process which transforms a descriptive ISO 9001 requirement into a prescriptive Quality Policy Statement.

We can clarify the language used in this graphic by a review of previous statements and definitions.

Policy

As defined by ISO 8402:1994(E/F/R):

➪ "Overall intentions and direction of an organization with regard to quality, as formally expressed by top management."

Policies are by their nature "time independent," that is, they do not describe movement but rather define position. Whereas procedures are "time-dependent," that is, they describe flow, process, and movement.

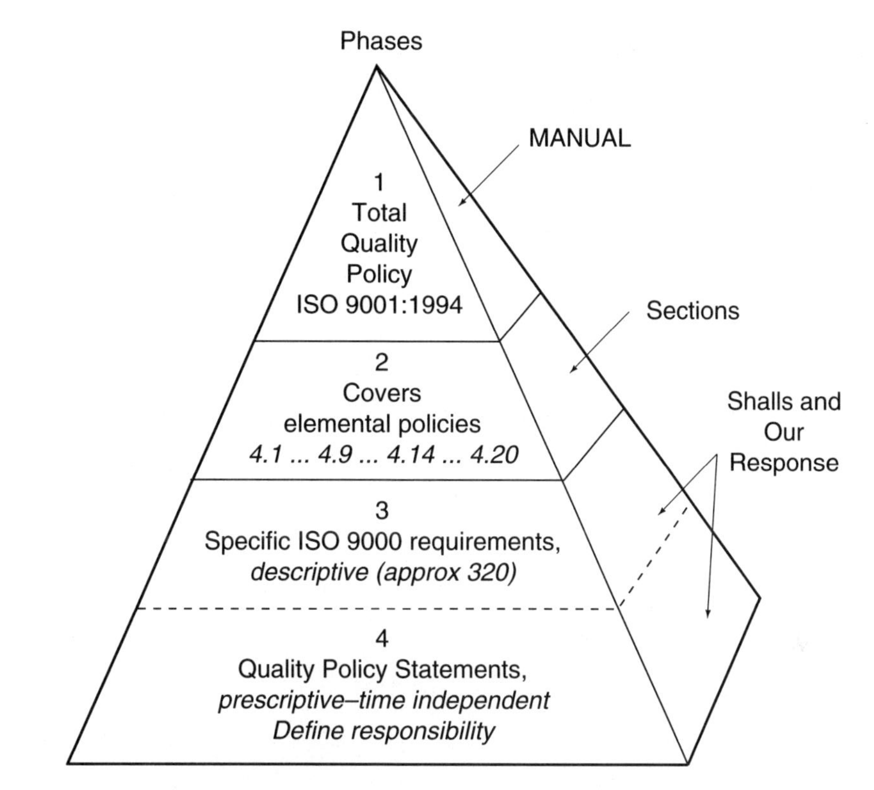

Figure 9.4 The four phases of the Quality Policy Manual. Direct sequence w/ANSI/ISO/ASQC Q9001-1994 sections.

A policy is basically "a rule of the house" set up by top management. Policies are *prescriptive* (specific direction) and indicate method of approach.

Total Quality Policy

The Total Quality Policy consists of all of the ISO 9001 quality policies presented in the Quality Policy Manual.

Elemental Policies

There are twenty (20) elemental policies in ANSI/ISO/ASQC Q9001-1994, which in the case chosen are usually presented in the Quality Manual in twenty (20) sections. More sections can be added, as necessary—for example, Security, Safety.

Specific ISO 9000 Requirements

Within each element of the Standard is a group of requirements defined by "shalls." As indicated in Section 2, Chapter 6, there are approximately three-hundred and twenty (320) explicit and implicit "shalls" in ANSI/ISO/ASQC Q9001-1994. These are stated in a *descriptive manner within the Standard.*

Quality Policy Statements

We have maintained that **each "shall" must be addressed** if the Manual is to clearly define the overall structure of the documented system and its effective implementation. We have shown in a previous section that the requirement for documentation that is effectively implemented is set forth in Clause 4.2.2 of the Standard.

- Thus, a Quality Policy Statement is required by the supplier in response to each "shall." The Quality Policy Statement is intended to be prescriptive and to delineate authority/responsibility.

Some examples of Quality Policy Statements are shown in Table 9.2 entitled "Examples of Quality Policy Statements." Appendix D, entitled "Further Examples of Quality Policy Statements," covers several more areas of the Standard not already addressed in the main text.

Notice how specific each statement is and how authority and responsibility are clearly stated. In just these few sentences it is possible to begin to visualize the Excellent Corporation's management structure and commitment to quality.

Shall Conclusion

- We conclude that unless each "shall" of the Standard is addressed clearly within the Manual—in the form of Quality Policy Statements—there is a high probability that key sections of the three ISO 9000 pillars, i.e., documentation, implementation, and demonstration of effectiveness, will be trivialized and undermine the integrity of the entire Standard.

Part 4: Information Flow via the Tiers

Waterfall Effect

Once the four-tier hierarchy has been established, the total documentation system tends to behave with a waterfall effect, that is, the number of process documents is less than the number of procedural documents, which in turn is less than the number of forms.

We have shown this effect graphically in Figure 9.5, entitled "ISO 9001 Documentation Waterfall Effect." In this figure, we have again made the assumption that the Quality Policy Manual is a stand-alone document. As we indicated previously, this is not the only possible configuration for the system, but it greatly helps to describe our concept.

The tendency for documentation growth must always be challenged. However, the use of the techniques described in this book will tend to minimize this growth.

TABLE 9.2 Examples of Quality Policy Statements		
ANSI/ISO/ASQC Q9001-1994 Clause	**ANSI/ISO/ASQC Q9001-1994 "Shall"**	**Quality Policy Statement**
4.1.1 Quality policy	The supplier shall ensure that this policy is understood, implemented, and maintained at all levels of the organization.	▪ The Excellent Corporation's Quality Policy Statement is posted throughout the facility and published in the monthly corporate bulletin. ▪ Each department manager holds a monthly training session with staff to review quality issues.
4.4.2 Design and development planning	The supplier shall prepare plans for each design and development activity.	▪ The project engineer and the program manager are responsible for the creation of a project time line. ▪ Project planning software is used to define every stage of the design process, and the plan is subject to the approval of the chief engineer.
4.6.2 Evaluation of sub-contractors	The supplier shall: establish and maintain quality records of acceptable subcontractors (see 4.16).	▪ The purchasing manager establishes and maintains the approved vendor list (AVL), which contains all Class A subcontractors, i.e., critical suppliers. ▪ Class B and C suppliers are used at the full discretion of the buyers.

ISO 9000 Hierarchal Drivers

We have also shown graphically in Figure 9.6, entitled "ISO 9001 Hierarchal Drivers," that the four-tier concept is universal, that is, the Standards define the Quality Manual responses, the Quality Manual responses confine the content of the second-tier documents, and the Tier II documents drive the content of the procedures/work instructions. In this manner, the executive rules are transformed into management controls, which are then transformed into operational directives.

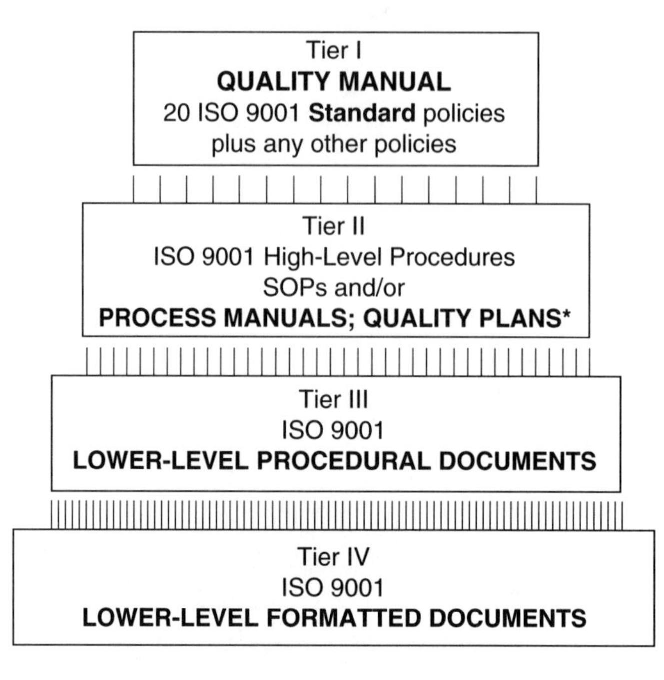

Figure 9.5 ISO 9001 documentation waterfall effect.

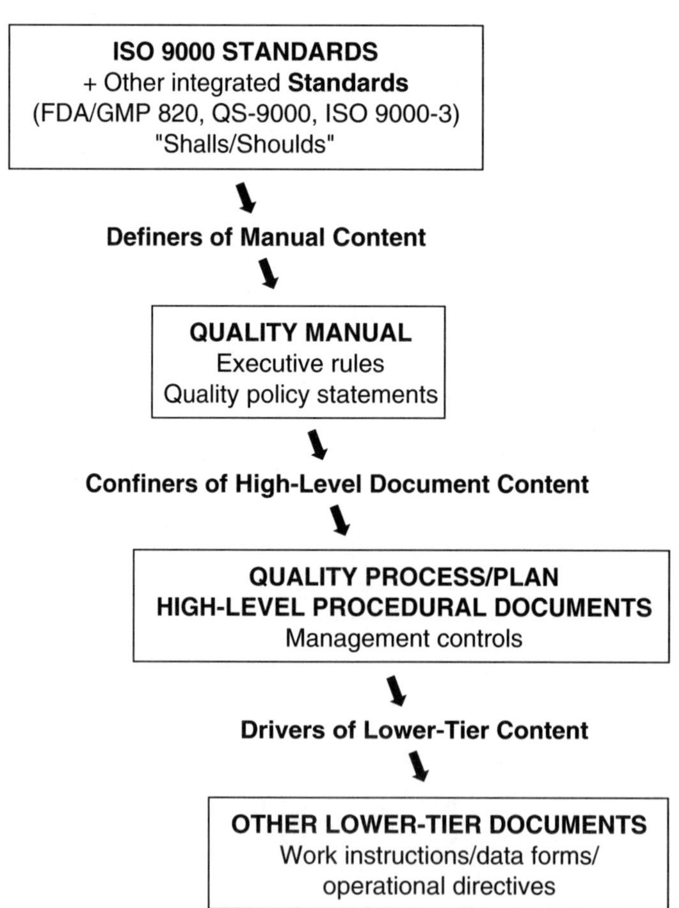

Figure 9.6 ISO 9000 hierarchal drivers.

Thus, the design of an effective QMS is more than the sum of its parts. Unlike the engineering design of a personal computer's printed wiring assembly, there is a powerful motivational element present within the QMS. For example, there is no need to motivate the electrons to flow efficiently within the printed circuit board's copper tracks, but there is an extremely important requirement to create a symbiotic relationship between the inert document's pages and the dynamic application of those documents by human beings.

The QMS acts as a living organism, and this is why it is so difficult to create it in the first place and then to effectively maintain the system. However, it is the inherent ability of the four-tier structure to enhance informational flow that increases the probability of a successful QMS.[2]

Endnotes

1. Although the terms Quality Policy, Process, and Procedure are defined in *ISO 8402:1994(E/F/R),* "Quality management and quaility assurance—Vocabulary," a serious schism exists in the application of these concepts. We have based our discussion on the work of Robert E. Horn, as exemplified in Demystifying ISO 9000, Second Edition, Information Mapping, Incorporated, Waltham, MA, 1994, pg. 5–6, for example.
2. The importance of human interfacing with the QMS is extremely well documented. A source of original and lucid studies in this matter is available in the *Quality Management Journal,* a publication of the ASQ, for example, Volume 4, Issue 2, 1997.

Chapter 10
Manual Sequences

Part 1: Three Possible Manual Sequences

The actual structure of the Manual depends upon the nature of the business and the manner in which we intend to propagate information within the system. There are at least three basic configurations for the Manual that are compliant with the ISO 9000 requirements (as long as the relationship to each ISO 9000 **Standard** is clearly defined):

- **Direct sequence** based on the ISO 9001 sequence, that is, 4.1, 4.2, 4.3, 4.4, 4.5,. . . 4.20
- **Shewhart Cycle sequence,** that is, Plan, Do, Check, Act discussed earlier
- **Operational Cycle sequence,** for example, Marketing & Sales, Production Control, Purchasing, Kitting, Assembly, Test, Shipping, Customer Service.

Part 2: Direct Sequences

The pertinent clauses ("shalls") of ANSI/ISO/ASQC Q9001-1994 consist of twenty sections numbered consecutively from 4.1 (Management Responsibility) to 4.20 (Statistical Techniques). It is very common to find Manuals configured in this fashion, that is, as sections that correspond directly with the numbering system of the Standard.

The specific sequences in the ISO 9000 contractual Standards are summarized in Table 10.1, entitled "ISO 9000:1994 Contractual Standards Comparison." We note that ISO 9002 is exactly the same as ISO 9001, except for element 4.4 Design Control. ISO 9003 is seldom applied, although it has been greatly enhanced over the 1987 version.

Assessments

ISO 9000 auditors commonly assume the existence of a directly sequenced Manual. Furthermore, they assume a one-to-one correspondence with each clause in the element. The form of the certification assessment generally takes on the

Table 10.1 **ISO 9000: 1994 Contractual Standards Comparison**			
Element No.	**ANSI/ISO/ASQC Q9001-1994 Title**	**ISO 9002:1994**	**ISO 9003:1994**
4.1	Management Responsibility		
4.2	Quality System		
4.3	Contract Review		
4.4	Design Control		
4.5	Document & Data Control		
4.6	Purchasing		
4.7	Control of Customer Supplied Product		
4.8	Product Identification & Traceability		
4.9	Process Control		
4.10	Inspection & Testing		
4.11	Control of Inspection, Measuring, & Test Equipment		
4.12	Inspection & Test Status		
4.13	Control of Nonconforming Product		
4.14	Corrective & Preventive Action		
4.15	Handling, Storage, Packaging, Preservation & Delivery		
4.16	Control of Quality Records		
4.17	Internal Quality Audits		
4.18	Training		
4.19	Servicing		
4.20	Statistical Techniques		

Legend

Comprehensive	Less than ISO 9001, or ISO 9002	Element Not Present

Source: Based on data from ANSI, *ISO 9000 International **Standards** for Quality Management*, 3rd ed., Geneva.

form shown in Table 10.2, entitled "ISO 9001 Certification Assessment for Excellent.[1]

Notice the way that certain elements are grouped, for example

- 4.3 Contract Review with 4.4 Design Control
- 4.6 Purchasing with 4.7 Control of Customer Supplied Product and 4.19 Servicing.

Such groups reflect the concomitant relationships of one Standard element to another.

TABLE 10.2

ISO 9001 Certification Assessment for Excellent

Time	Activities	Assessors	Excellent Guides
Day One			
8:15 AM	Arrive on site	Team	Core Grp
8:30	Opening meeting chaired by the lead assessor	Team	Core Grp
9:00	Tour of the plant	Team	Mngt Rep
9:30	4.1 Management Responsibility; 4.17 Internal Quality Audits; 4.18 Training	Team	Core Grp
Noon	Lunch	Team	Core Grp
1:00 PM	4.2 Quality System—Standards & Codes, Factored Items 4.9 Process Control—Preventive Maintenance	Lead Assessor	Core Grp Jay S.
2:00	4.5 Document & Data Control; 4.14 Corrective & Preventive Action & Customer Complaints 4.8 Product Identification & Traceability; 4.12 Inspection & Test Status	Lead Assessor	Fran L. Jay S.
3:00	4.6 Purchasing; 4.7 Control of Customer Supplied Product; 4.19 Servicing 4.10 Inspection & Testing	Lead Assessor	David W. Judy S.
4:00	Prepare for verbal feedback	Team	
4:30	Verbal Feedback chaired by lead assessor	Team	Core Grp
5:00	Exit site	Team	

Table 10.2
Continued

Time	Activities	Assessors	Excellent Guides
Day Two			
8:30 AM	Review of Previous Day's Findings	Team	Core Grp
9:00	4.3 Contract Review; 4.4 Design Control 4.9 Process Control continued	Lead Assessor	David W. Jay S.
10:00	4.13 Control of Nonconforming Product 4.15 Handling, Storage, Packaging, Preservation, & Delivery	Lead Assessor	Judy S. Mark J.
11:00	4.20 Statistical Techniques; 4.16 Control of Quality Records 4.11 Control of Inspection, Measuring, & Test Equipment	Lead Assessor	Mark J. Fran L.
12:00	Lunch	Team	Core Grp
12:30	Prepare for verbal feedback	Team	
1:00 PM	Verbal feedback chaired by lead assessor	Team	Core Grp
1:15	Assessors prepare for the closing meeting	Team	
4:30	Closing meeting chaired by lead assessor	Team	Core Grp

Registrars and third-party assessors tend to gravitate toward this type of element alignment because it is an efficient way to do the audits against very stringent time constraints. In a certification audit, the assessor must keep moving along and cover all of the elements within the time frames noted in the far left column. There is very little opportunity to go back and check out an observation later on in the audit.

Table 10.3, entitled "ANSI/ISO/ASQC Q9001-1994 Most Common Nonconformances," illustrates the general distribution of nonconformance reports in the initial assessment and the obvious and continued difficulties in document and data control. Elements 4.9, 4.10, and 4.11 are a poor second.

Checklist

To prepare the organization for this extensive audit, if the direct sequence method is chosen, a convenient checklist can be generated to track Manual progress. An

TABLE 10.3

ANSI/ISO/ASQC Q9001: 1994 Most Common Nonconformances

Element	Title	Percent of Total
4.5	Document & Data Control	17.0
4.9	Process Control	10.0
4.10	Inspection & Testing	10.0
4.11	Control of Inspection, Measuring, & Test Equip	10.0
4.6	Purchasing	8.0
4.1	Management Responsibility	7.0
4.3	Contract Review	6.0
4.18	Training	6.0
4.4	Design Control	4.0
4.13	Control of Nonconforming Product	4.0
4.17	Internal Quality Audits	4.0
4.14	Corrective & Preventive Action	3.0
4.15	Handling, Storage, Packaging, Preservation & Del	3.0
4.2	Quality System	2.0
4.8	Product Identification & Traceability	2.0
4.16	Control of Quality Records	2.0
4.7	Control of Customer-Supplied Product	1.0
4.12	Inspection & Test Status	1.0
4.19	Servicing	1.0
4.20	Statistical Techniques	1.0

Re: Based on data from QSU, p. 17, October 1995. Summation does not equal 100% due to rounding.

example of this checklist is shown in Table 10.4, entitled "Tier I ISO 9000 Quality Policy Manual Summary." We indicate the use of such a chart for elements 4.1, 4.2, 4.3, and 4.4 of the Standard. A similar set of checks can be generated for each of the elements.

As a rule of thumb, when a section is rated at 90 percent or higher, it is ready for the initial assessment.[2] The "OKs" indicate that every "shall" has been addressed and the "flags" indicate just the opposite. Either a specific action item is required to close the flag or a specific nonconformance (NCR) has been written against the section during an internal quality audit. There are five open NCRs indicated.

TABLE 10.4

Tier I ISO 9000 Quality Policy Manual Summary

Ref#:97-005-01 Supplier Name/Location: Type:
Dates: March 98 The Excellent
Corporation, Lexington, MA ISO 9001:1994

Sequenced Manual Section	Activity Covers Excellent's Quality Manual	Found in Text	Non-Confrm Issues	Comments
One	Management Responsibility Scope Quality policy Quality objectives Quality commitment Customer expectations & needs Organization—R&A Organization—Resources Management representative Management review	~90% OK OK OK OK OK OK OK OK OK	All resolved	Looks real good. Needs a final edit. One more mgmt review. Mgmt rep memo.
Two	Quality System Quality Manual Quality system procedures Quality planning Factored items Interdivisional interfaces	~80% Flag Flag OK OK Flag	NCR008 open	Needs interface clean-up. Not clear on Online system.
Three	Contract Review Review Amendment to contract Records	~85% Flag OK OK	NCR003 open	Clean up lower tier references.
Four	Design Control Design & development planning Organizational & technical interfaces Design input Design review Design output Design verification Design validation Design changes	~60% Flag OK OK OK Flag Flag Flag Flag	NCR001 open NCR012 open NCR015 open	.Q Policy not real yet. Needs work and a design control manual.

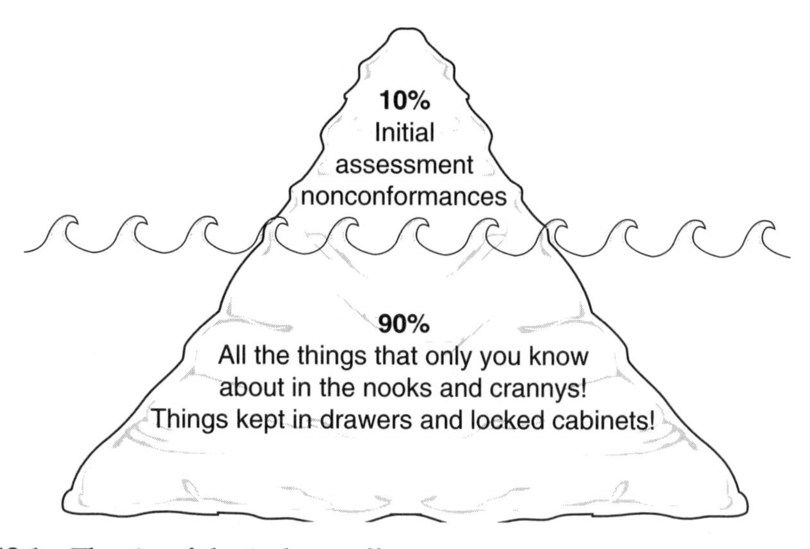

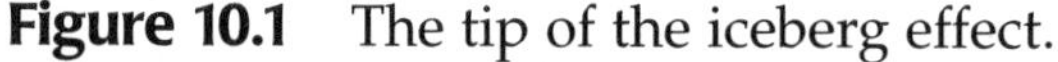

Figure 10.1 The tip of the iceberg effect.

Tip of the Iceberg

When the day of the initial assessment arrives, it is important to realize that the assessor's observations represent the tip of the iceberg (refer to Figure 10.1). They only see what they need to see in order to ensure themselves that the supplier has a workable QMS that will most likely produce a reasonable payback in a reasonable time. At least 90 percent of the nonconformances lie below the surface.

You, of course, know exactly what the nonconformances are, and the assessors rely on you to make those corrections as part of an effective QMS program. It is not uncommon to feel that you have put one over on the assessors once they leave. On the contrary, if you have, it is really cutting off your nose to spite your face. They saw it, but did not have the time to investigate. On the other hand, you know it is there. So you need to fix it.

Dynamics of the Initial Assessment

At the close of the initial assessment (IA), the lead assessor *recommends* either with or without condition. The Registrar's Executive Board *approves*. The several possible conditions for approval include (these vary considerably from registrar to registrar)

- All NCRs cleared during IA—recommend certification without condition
- Minors left to be cleared after IA—recommend certification but hold issuance until all are cleared
- Clear by plan to be followed up at first surveillance
- Some minors declared as "concerns" and to be monitored at the first surveillance
- Opportunities for Improvement—potential economic savings

The exception is in regard to *major* nonconformances. They are usually treated as follows:

- Majors left to be cleared during IA—requires a return audit of those areas within usually 90 days, then recommendation to certify[3]
- However, majors can be downgraded during the IA to avoid this problem.

The Essence of ISO 9000

It is abundantly clear during the initial assessment that the essence of ISO 9000 is to state with great clarity who manages, performs, and verifies the activities and functions of documentation, implementation, and demonstration of effectiveness. Such activities are normally illustrated through management reviews, internal quality audits, quality records, and corrective and preventive actions programs.

The point we want to make here is that you cannot fail an initial assessment, unless your simply quit. The worst thing that can happen is that it might take longer and cost more.

PART 3: SHEWHART SEQUENCE

The Shewhart Cycle of continuous improvement, that is, Plan-Do-Check-Act, can also be used to configure the Manual.

This is accomplished by a configuration that encapsulates the several elements of the Standard into sections defined as Plan-Do-Check-Act as previously demonstrated (refer to Section 1, Chapter 2). Previously we have never observed this method in action as part of an ISO 9000 assessment but we have now incorporated the concept into a combined TQM/ISO program which is designed for certification.[4]

In this approach, the relationship between the Standard's elements and the Shewhart Cycle varies with the TQM approach. Table 10.5, entitled "Quality Manual Contents—According to the TQM Model Employed—w/Related ANSI/ISO/ASQC Q9001-1994 Elements," demonstrates this method.[5] For example, in the PLAN cycle, the TQM model includes goals, marketing, estimating, and supplemental control.

By a directly integrated TQM/ISO program from the project start it is possible to gain the full advantage of both quality management concepts. We will cover this topic no further because there is a plethora of books available on this subject.[6]

PART 4: OPERATIONAL SEQUENCE

Another method is to lay out the Manual in terms of the organization's actual manufacturing or service processes. We have seen several attempts made at this approach but they were eventually rejected due to the correlation difficulty with the Standard's sections. However, as illustrated in Figure 10.2, entitled "The ISO 9001 Quality System Value Chain," the Standard's set of twenty sections does lend itself to an operational flow.[7]

TABLE 10.5

Quality Manual Contents—According to the TQM Model Employed—with Related ANSI/ISO/ASQC Q9001:1994 Elements

PLAN Cycle—Goals/Marketing & Sales/Estimating & Supplemental Control
- 4.1 Management responsibility
- 4.2 Quality system
- 4.3 Contract review
- 4.4 Design control
- 4.5 Document and data control
- 4.16 Control of quality records

DO Cycle—Purchasing & Project Coordination/Manufacturing
- 4.6 Purchasing
- 4.7 Control of customer-supplied product
- 4.8 Product identification and traceability
- 4.9 Process control
- 4.10 Inspection and testing
- 4.11 Control of inspection, measuring and test equipment
- 4.12 Inspection and test status
- 4.15 Handling, storage, packaging, preservation and delivery

CHECK Cycle—Job Costing & Cash Flow Control
- 4.17 Internal quality audits
- 4.19 Servicing
- 4.20 Statistical techniques

ACT Cycle—Financial Feedback & Cost of Quality
- 4.13 Control of nonconforming product
- 4.14 Corrective and preventive action w/customer complaints
- 4.18 Training

A compromise position is attained between a true sequence and an operational sequence by means of a Tier II, "Manufacturing Process Manual." This type of document is essentially a quality plan, since it begins with contract review and flows through to servicing. Its use permits the reader to more readily sense the operational flow as the Quality Policy Manual is read.

The "Manufacturing Process Manual" is generally created with flow charts and supplementary text. The elements covered by the "Manual" are 4.3, 4.6, 4.7, 4.8, 4.9, 4.12, 4.15, and 4.19. When flow charts are used, the entire manufacturing process can be posted in key areas of the facility, and it becomes an impressive display for visitors.

Value Chain

As mentioned previously, Figure 10.2 demonstrates this operational flow inherent in the Standard, as indicated by the central flow of activities from 4.3 Contract Review to 4.19 Servicing. This intrinsic operational flow is supported by a set of executive functions, that is, 4.1 Management Responsibility, and control functions, that is, 4.4 Design Control.

Executive functions →

EXEC
4.1 Management Responsibility & 4.2 Quality System
4.6 Purchasing
4.14 Corrective & Preventative Action & Customer Complaints
4.17 Internal Quality Audits
4.18 Training
4.20 Statistical Techniques

↓ ↓

Contract Review 4.3 Mktg & Sales →	Rcvg Stocking Inspection 4.10 4.15 →	4.9 Process Control 4.8 Product ID & Tr 4.10 Inspect & Test 4.12 Insp & Tst Status →	Delivery Warehouse 4.15 HSPPD →	Ship	Service Install Repair 4.19 Svcg →	C/I Cycle ↑ →

↑ ↑

CONTRL
4.4 Design Control
4.5 Document & Data Control
4.7 Control of Customer Supplied Product
4.11 Control of Inspection, Measuring, & Test Equipment
4.13 Control of Non conforming Product
4.16 Control of Quality Records

Control functions →

Figure 10.2 The ISO 9001 quality system value chain.

A similar diagram can be constructed for both the subcontractor and the customer. We can indicate the role of the supplier (certified organization/you) when the value chain is extended to include this complete interorganizational flow. This unique functionality can be demonstrated as follows (each arrow represents the set of executive functions, operational process, and control functions shown in Figure 10.2).

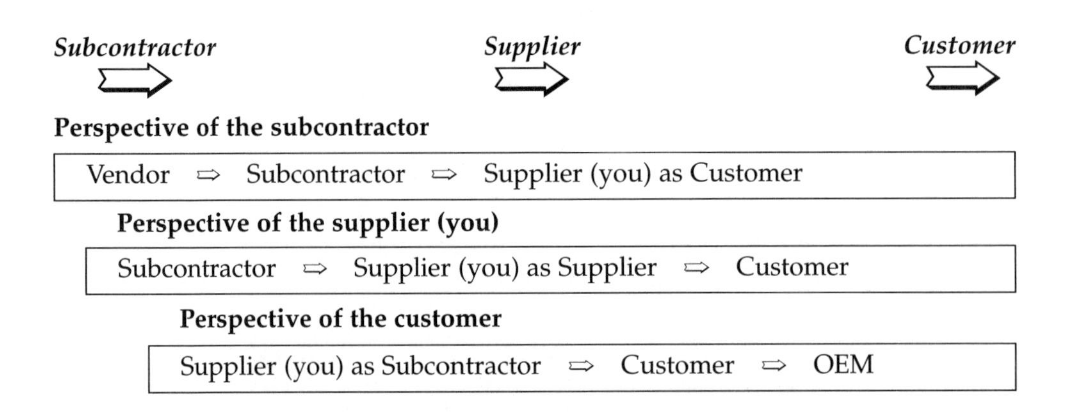

As indicated above, the supplier fulfills the role of customer, supplier, and subcontractor dependent upon where its various transactions occur in the chain. The middle diagram—Perspective of the supplier (you)—represents the basic terminology used in the ANSI/ISO/ASQC Q9001-1994 Standard, that is, the certified organization is always termed the "supplier."

In practice, this nomenclature results in confusion since it is common to call subcontractors "suppliers." It is necessary only to define your terms in ISO 9000 to resolve this issue.

A typical set of alternatives in this regard is shown in Table 10.6, entitled "Supply Chain Interfaces." The terminology used is related to the several ISO Standards and guidelines that cover this topic.

TABLE 10.6

Supply Chain Interfaces*

(Re: ISO 9000-1:1994(E) & ISO 8402:1994(E/F/R))

Subcontractor ⇨	**Supplier ⇨**	**Customer**
ISO 9001 ISO 9002 ISO 9003	ISO 9001 ISO 9002 ISO 9003	ISO 9001 ISO 9002 ISO 9003
Subsupplier ISO 9000-1	***Supplier or organization*** ISO 9000-1	***Customer*** ISO 9000-1
Subcontractor ISO 9004-1	***Organization*** ISO 9004-1	***Customer*** ISO 9004-1
Vendor ***Who supplies you***	***Contractor*** ***Certified organization*** (you) ***Business First Party***	***Purchaser*** ***Whom you sell to*** ***Business Second Party***
Metrology house, raw materials producer, consultants, contract supplier, design house, component manufacturer, private label manufacturer	Producer, distributor, importer, assembler, service organization, software house, manufacturer, design house (may be internal or external to organization)	Independent supplier and producer, sister division, joint venture partner (may be internal or external to organization)

*The ISO 9000 Technical Committee, TC 176 hopes to harmonize this terminology in the Year 2000 revision of the standards, e.g., "Writing Under Way on Next ISO 9000 Standards," *Quality Systems Update,* December 1997, Vol. 7, No. 12, The McGraw-Hill Companies.

Part 5: Comparison of Sequences

The comparative analysis for the three sequences is shown in Figure 10.3, entitled "Comparison of Quality Policy Manual Content Attributes."

What we observe is that the Direct Sequence is the most efficient form of linkage to the lower-tier documents and is the most common sequence used. It does require a minor search to find the critical operational functions and the continuous improvement cycle.

The Shewhart Cycle could become more popular as ISO 14000 becomes more widely known. However, unless ISO 9001 is recrafted into the Shewhart format, the issues of linkage and operational overview will remain.

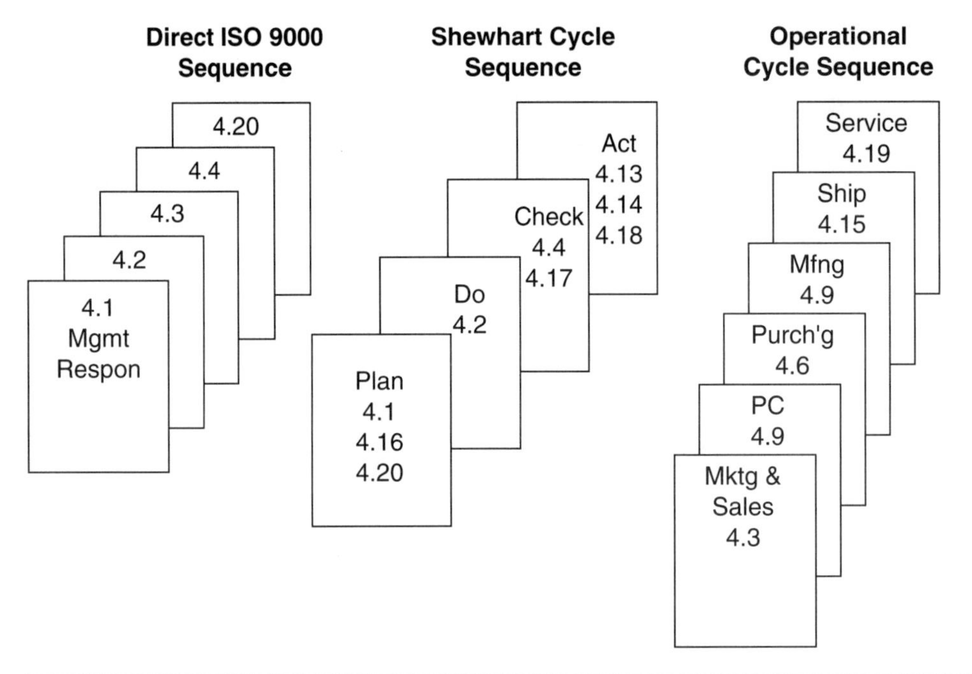

Attributes	Direct Sequence	Shewhart Cycle	Operational
Linkage to ISO Clauses	**Direct**	Need to search out	*Obscured*
Operational Overview	Need to search out	*Obscured*	**Direct**
Continuous Improvement Cycle	Need to search out	**Direct**	*Obscured*
Overall	**Most common**	*Have not seen used in ISO 9000, but adopted in ISO 14000 & initial use in TQM/ISO programs*	Least common

Figure 10.3 Comparision of Quality Policy Manual content attributes.

The Operational Cycle has both the linkage and continuous improvement cycle issues to overcome. As a result, we believe that the Direct Sequence method will continue as the prime approach.

Endnotes

1. For more information on the way a registrar views a client see, for example, *Meet the Registrar: Firsthand Accounts of ISO 9000 Success from the Registration Source*, by C. Michael Taylor, ASQ publications, Item: H0967, ISBN 0-87389-423-3, 1997.
2. The "90 percent rule is not meant to negate the desire for perfection. It does emphasize the reality of life among those who already have a full-time job in addition to their ISO functions. At any time, prior to certification, during certification, and after certification, there will always be about 10 to 20 percent of the system under revision. That is called "continuous improvement." Any attempt to combat or change this fact of life will be met with extreme frustration, followed by rebellion. Minor nonconformities due to "fine-tuning" are a constant-of-the-motion in ISO.
3. Although it is possible to have the registrar declare the organization "noncertifiable", we know of no such case in the hundreds of certifications with which we are familiar. The only situation under which this might occur is if the facility has obvious safety and/or hazardous waste nonconformances so that the assessors cannot perform their audit in a safe manner.
4. The recent pulication of the ANSI/ISO 14001–1996, "Environmental Management Systems—Specification with Guidance for Use" document is based upon the Shewhart Cycle. There is a general interest in reconfiguring ISO 9001 to conform to this approach.
5. Based on the work of Dr. Anthony F. Costonis, President & Founder of Corporate Development Services, Inc. (CDS), of Lynnfield, MA, *HTTP://WWW.Corpdevelopment.com.*
6. See for example, the latest ASQ Quality Press Catalog, p. 2, Fall/Winter 1997.
7. The concept of a value chain was first introduced by Michael E. Porter in a series of business textbooks, for example, *Competitive Advantage*, The Free Press, New York, 1985. The idea of "chains" finds many outlets, for example, "The Quality Chain", Tom Troczynski, *Quality Progress*, September 1996, p. 208.

CHAPTER 11
MANUAL CONFIGURATIONS

PART 1: TWO UNIQUE CONFIGURATIONS

We will now demonstrate that regardless of which model is chosen for the Manual's section sequences—Direct ISO Sequences, Shewhart Cycle, or Operational Cycle—there are effectively only two unique ways to design the Manual's configuration, either as a

- Model I—Stand-alone document that deals only with policy
- Model II—Integrated policy/Tier II document that contains both policy and procedure

We maintain that any other form of documentation is a variation of one of these two basic forms. In addition, we will treat the manner in which linkage from the Manual to lower-level documentation can be performed effectively.

- The configuration decision is paramount in the choice of just where to place the Quality Policy Statements. We have found it to be the primary source of conflict in Manual design.

PART 2: THE STAND-ALONE CONFIGURATION—MODEL I

In the case of a stand-alone Manual—if the writer meticulously follows the criteria stated in the previous sections for the structure of a quality policy statement—there will be no policy statements in the lower-tier documents, because it will be unnecessary.

The stand-alone Manual clearly references each lower-tier document that it directly affects; for example the Manual's "Section 14 Corrective & Preventive Action & Customer Complaints" would send the reader to an SOP entitled "Corrective Action Procedures."

This requirement is based upon Par *4.2.1 Quality System-General* of the Standard,

➪ "The quality manual shall include or make reference to the quality-system procedures . . . ."

Section References

This is normally done by inserting a reference document in each section of the Standard. Such an approach can produce a very complex and difficult navigational problem if there are many Tier II documents. A typical example of this complex case would be in "Section 9 Process Control," which could look like

➪ "The Excellent Corporation documents related to process control are to be found under:
   1. Doc# 2-09-012, Production Control Procedures
   2. Doc# 2-09-011, Materials Control Procedures
   3. Doc# 2-09-004, Use of the Traveler Procedure
      . . . .
      . . . .
   12. Doc# 2-09-008, Release of Capital Equipment."

- A way to avoid this difficulty is to create a process-related "Hub document," which acts as a documentation flow center.

Hub Documents

Definition—A "Hub document" is similar to an airport hub in that it is a center of information flow. The Manual need only reference this one document in each section, and the Hub document will then take the reader to the appropriate supplementary lower-tier documents.

For example, in a structured, hypertext system, that is, on-line, there would be only one icon per section of the Standard. Then, once you are in that referenced document, other icons would transfer you to the appropriate supplemental document.

It is unnecessary to have a hypertext system to structure a "hub" system. This degree of simplicity is available regardless of the word processor design.

Following, we offer an example of how a system can be simplified and made more user-friendly without compromising the system's basic integrity.

Hub Template—A typical master reference list of a "Hub document"–oriented Manual is shown in Table 11.1, entitled "Typical Hub Document References." Each section of the Manual would refer to the appropriate hub document. An alternative method of display would be, for example, by means of a documentation tree.

Some key attributes of the hub design are

- In most cases, there is only one Hub document.
- Several hub documents are used where they either represent unique processes or require a summary statement, for example, business plan, response to 4.2.
- A manufacturing process document, element 4.9, is used as a hub to avoid redundancy with concomitant elements, for example, 4.8, 4.12, 4.13, and 4.15.
- Hub documents are invariably very high-level process or procedural documents.

TABLE 11.1

Typical Hub Document References

ISO 9001 Element	Manual Section	Typical Hub Documents	Typical Contents
4.1	One	▪ Business Plan ▪ Security Manual ▪ Employee Manual	Very high-level policy documents authored by top executives
4.2	Two	▪ Quality Policy Manual ▪ Process Master Manual ▪ Procedure Master Manual ▪ Forms Master Manual	Very high-level operational documents authored by top executives
4.3	Three	▪ Sales & Marketing Manual	All aspects of contract review and marketing
4.4	Four	▪ Design Control Guidelines	Hardware, software, CAD processes
4.5	Five	▪ Document Control System	QA, engineering, MRP processes
4.6	Six	▪ Purchasing Manual	AVL, vendor evaluation, P.O.s
4.7	Seven	▪ Customer-Supplied Material Procedure	RMAs, materials, capital equipment
4.8	Eight	▪ Manufacturing Processes	Integrated into manufacturing phase
4.9	Nine	▪ Manufacturing Processes	Integrated approach to the total manufacturing process
4.10	Ten	▪ Inspection & Test System	Receiving, in-process, final, assembly, test
4.11	Eleven	▪ Metrology Manual	In-house and subcontracted calibration practices
4.12	Twelve	▪ Manufacturing Processes	Integrated into manufacturing phase
4.13	Thirteen	▪ Manufacturing Processes	Integrated into manufacturing phase
4.14	Fourteen	▪ Corrective & Preventive Actions & Customer Complaints	Integrated approach to corrective action, preventive action, and customer complaints
4.15	Fifteen	▪ Manufacturing Processes	Integrated into manufacturing phase

Table 11.1 Continued			
ISO 9001 Element	**Manual Section**	**Typical Hub Documents**	**Typical Contents**
4.16	Sixteen	▪ Records Manual	Covers QA, engineering, manufacturing records
4.17	Seventeen	▪ Quality Audit Manual	Contains internal quality, vendor, customer, and third-party audit protocols
4.18	Eighteen	▪ Training Manual	Manager, auditor, OJT training protocols
4.19	Nineteen	▪ Customer Service Manual	On-site service, warranty protocols
4.20	Twenty	▪ Statistical Analysis Manual	Sampling plans, Pareto charting, SPC, DOE

Direct Sequence Is the Most Common Sequence for a Stand-Alone

The exact form of the stand-alone Manual has been almost entirely based on the direct ISO 9000 sequence, as described previously, that is, a one-to-one correspondence to the numbering system of the Standard.

Quality Policy Statement Imperative

By definition, as defined in ISO 8402:1994(E/F/R), Par 3.12, a quality manual is a

➪ "document stating the quality policy and describing the quality system of an organization."

- Therefore, once the supplier has designated a stand-alone document as the "Quality Manual"—regardless of the way a supplier chooses to label the contents of that Manual—one thing we believe is clear due to its definition is that all Quality Policy Statements, to whatever level of detail is required, should be contained in this document.

Primary Source of Inconsistency

The Stand-alone Manual configuration is the primary source of inconsistency in Manual design because of the tendency to place policy statements not only in the Manual but in lower-level documents as well.

Application to Third-Party Assessments

It is important to keep in mind that third-party assessors tend to go "shall" by "shall" through the Manual on each clause, and each "shall" must be addressed. Although many of the Quality Manual nonconformances appear to be of a minor nature—there is an assumption that the actual process exists (if not, it could be a major)—they would still require nonconformance reports (NCRs) and time to respond.

In some cases, for example, for a rewrite of 4.10 Inspection & Testing, the amount of rewrite is so extensive that it will probably have to be done through the mail. This is certainly not the end of the world, but it could delay the certification by some number of weeks.

Corrective Actions for Dislocated Quality Policy Statements

If, as is usually the case, the quality policy statements are located in lower-tier documents, any one of the following corrective actions has been found to be acceptable:

- Cut and paste the statements into the Quality Manual (this minimizes redundancy).
- Copy the statements into the Quality Manual (creates redundancy but is not a nonconformance unless the lower-tier document statements disagree).
- Include the lower-tier document as part of the Quality Manual (this makes for a very heavy Quality Manual but is acceptable).

From an operational standpoint, although the placement of Quality Policy Statements into lower-tier documents instead of in the Quality Manual is commonplace, the fact that they are in a lower-tier document that is not part of the Quality Manual is unacceptable because

- Decision makers who require quality policy information would seldom (if ever) have the whole set of lower-tier documents at their disposal.
- Lower-tier documents generally contain proprietary information, and are restricted in their distribution.
- This is why essentially all suppliers have chosen the "stand-alone" form of the Quality Manual.

Guidance on this subject is noted in the International Standard (refer to ISO 10013, "Guidelines for developing quality manuals").

PART 3: THE INTEGRATED MANUAL CONFIGURATION

In an integrated Manual the Quality Policy Statements and the lower-tier documents—especially Tier II—all appear in the same source. This is an approach that was commonly used some thirty plus years ago and can still be found in Mil-Q-9858A and FDA/CGMP-oriented manufacturing operations, as well as in the automotive industry.

In practice, in smaller companies, the Manual and the set of standard operating procedures (SOPs) are often distributed together in response to the Standard's element 4.5 Document and Data Control requirement that

⇨ "the pertinent issue of appropriate documents are available at all locations where operations essential to the effective functioning of the quality system are performed. . . . "

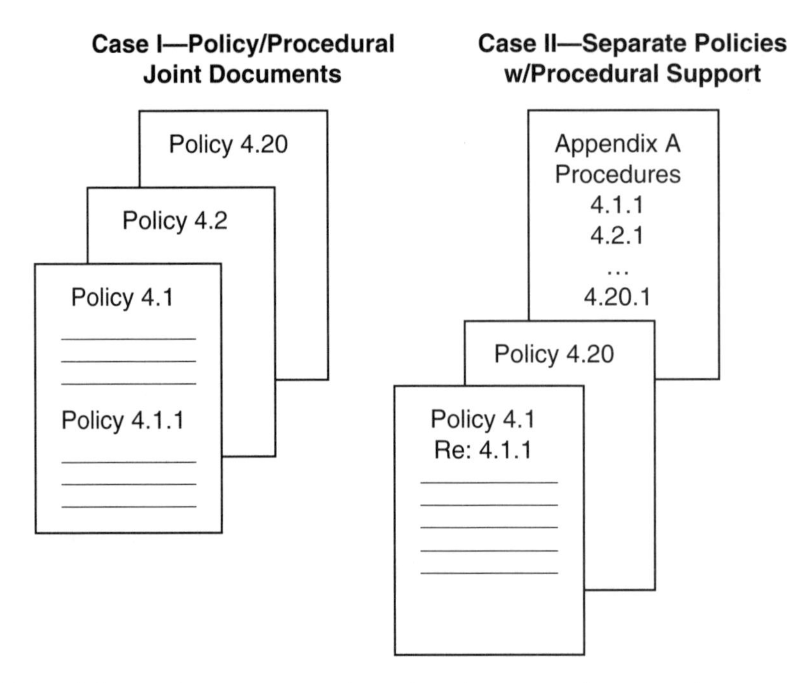

Figure 11.1 Schematic representation of an integrated manual.

Let us consider the implications of the integrated Manual approach.[1]

Specifically, let us assume that we have chosen the integrated Quality Manual configuration, that is, the Manual will contain all of the Quality Policy Statements required by the Standard in the form of either

- Case I—SOPs that begin with a policy statement and are immediately followed by the procedural information (a joint document)
- Case II—Abbreviated Quality Policy Statements that reference an attachment "A" that contains pertinent Tier II procedures (SOPs) as well as the rest of the required Quality Policy Statements.

Figure 11.1, entitled "Schematic Representation of an Integrated Manual," illustrates the form that such Manuals would take.

As we noted earlier, such an approach is in full compliance with the Standard, that is, clause 4.2.1 states that

⇨ ". . . The quality manual shall include or make reference to the quality-system procedures. . . ."

Implementation

However, the implementation of such a decision is far from trivial, and we want to outline some of the more obvious issues:

- When the assessors ask for the Manual, they will expect to see the complete Manual, that is, either the full joint policy and Tier II documented Manual or all of the front-end text and the SOP Appendix.

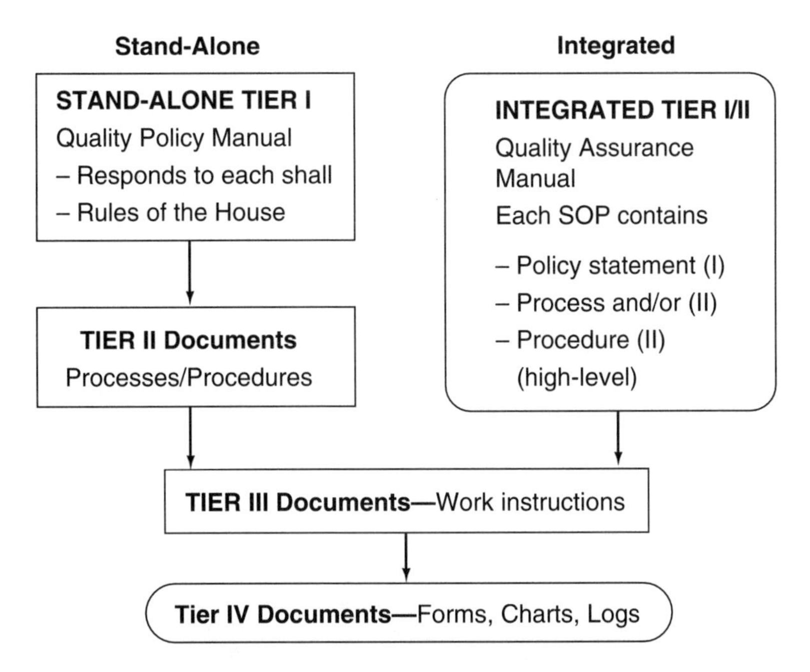

Figure 11.2 Comparative analysis of configurations.

The Appendix must include the full set of SOPs that provide the rest of the required quality policy statements.

- In the case of an Appendix, assessors will expect to see the SOP attachment fully noted as part of the Manual.
- All attachments will need to be under document control, however, the list of SOPs in the Appendix need not show the revision dates since they will not expect you to keep more than one master list of documents. But they will need to know the full contents of any appendices, either up front in the table of contents or as a part of the attachment.
- When assessors check the distribution of controlled Manuals they will expect to see a complete document in the auditee's area. The "complete document" is the same as defined above, that is, either the full jointly documented Manual or the Policy/Appendix SOP Manual.
- When a specific Quality Policy Statement is searched for—in response to a "shall" in the Standard—they can accept its presence anywhere in the integrated Manual, that is, front-end text or attachments.
- Assessors will expect to see a defining statement in Section 2 Quality System of the Manual to the effect that "the Manual does consist of a set of either jointly documented policy/procedural text or of abbreviated Quality Policy Statements and associated 'SOPs' contained in an Appendix."

Figure 11.2, entitled "Comparative Analysis of Configurations," offers a qualitative comparison of the stand-alone versus the integrated Manual approach. The comparative analysis is indicated in Table 11.2.

Table 11.2 (for Figure 11.2)
Comparative Analysis of Configurations

Attributes	Stand-Alone Manual	Integrated Manual
Size Scales with Org. size and complexity.	30 to 75 pages w/o appendices Only Quality Policy Statements	Hub approach—200 plus pages w/o appendices & forms; can be considerably bigger without a hub approach or for multi-site Manuals; only Tier I/II documents for a single division or site
Redundancy	May have some with Tier II	Low probability
Linkage	By reference—somewhat difficult to learn	Direct linkage
As a teaching tool	Powerful with a big picture viewpoint	Somewhat limited because of narrower viewpoint
As a marketing tool	Outstanding—can clearly define organization's personality and image	Not really viable for marketing purposes; an issue of proprietary information and big picture focus
Ease of auditing	Readily available information; Quality Policy Statements readily found	Readily available information; Quality Policy Statements could be obscured
Ease of distribution	Can distribute Manuals & SOPs to custom fit the area	Each SOP area requires the entire Manual
Overall	Most common in use	Not commonly found to date in ISO 9000 certifications

Discussion

As a result, the true difficulties with the integrated Manual approach are in regard to

- Size
- Marketing purposes
- Distribution

This approach could get even more cumbersome in a multidivisional operation where it might happen that one Manual was to be used in all divisions, with the result that not only the corporate-level procedures but also the divisional level procedures would be combined in one binder.

We have observed organizations' attempts to use an integrated Manual approach by modifying their present Mil-Q-9858A documentation, but they have eventually changed to the stand-alone model for ease of use and distribution. However, this approach has been used successfully in those cases where the supplier had little concern in regard to the proprietary nature of the Tier II documents, and thus was not concerned about their distribution as part of a Manual.

Part 4: Multidivisional Manuals

Expansion

The previous discussion can be readily expanded to include very large organizations with multidivisional requirements. Regardless of which basic configuration is chosen, the divisional documents would follow directly, the Corporate Manual would deal with policy at the highest level and the Divisional Manuals would respond to each corporate policy according to the specific operational characteristics of the respective division.

The Corporate Manual references the Divisional Manuals as appropriate.

Some examples of the content of a Divisional Manual versus the Corporate Manual are summarized in Table 11.3, entitled "Comparison of Corporate versus Divisional Manuals."

Table 11.3

Comparison of Corporate versus Divisional Manuals

Subject	Corporate Manual	Divisional Manual
4.1 Management Reviews	▪ Discusses the quarterly management review held at corporate with divisional managers present. ▪ Describes the various corporate feeder meetings that take place prior to the quarterly.	▪ Discusses the monthly divisional management review, which feeds the corporate quarterly review. ▪ Also discusses the divisional feeder meetings held locally prior to the monthly.
4.2 System Documentation	▪ Describes the entire quality system based on corporate standard operating procedures (CSOPs) and/or corporate process documents (CPDs). ▪ Discusses thevarious ways that the divisions interface with both the corporate office and other divisions.	▪ Descrbies the response to the CSOPs via divisional work instructions (DWIs). ▪ Describes specific interface functions with corporate and other divisions. ▪ Divisional SOPs are optional, and are generally redundant.
4.3 Contract Review	▪ Describes the highest level contract review policies. ▪ Example—"At the Excellent Corporation, all RFQs are reviewed and approved by the president."	▪ Describes the method used at the divisional level to meet the corporate policies. ▪ Example—"Divisional RFQ review and approval is by the general manager unless subject to corporate approval based on price."

Divisional

A schematic of such a multidivisional Manual structure, which follows the direct sequence pattern of ISO 9001 sections, is shown in Figure 11.3, entitled "Possible Configuration for a Corporate/Divisional Integrated Manual.

ISO Management Representative Example of Labels

The block labeled "C 4.1 Mgmt Rev" represents the corporate response to the "shalls" in element 4.1, Management Review, of the Standard, for example

- Corporate Memo—"The President has designated the corporate Director of Quality Assurance as the ISO 9000 management representative. A memo to this effect, which defines the additional duties and authority of this position, has been released by the President of the corporation and distributed to all employees of the corporation."

The block labeled "D 4.1 Mgmt Rev" represents the Divisional response to the Corporate Manual, for example

- Division Memo—"The General Manager has designated the division's Manager of Quality Assurance as the Site ISO 9000 management representative. Notice to all divisional employees was by means of the weekly quality improvement reviews held with the division's General

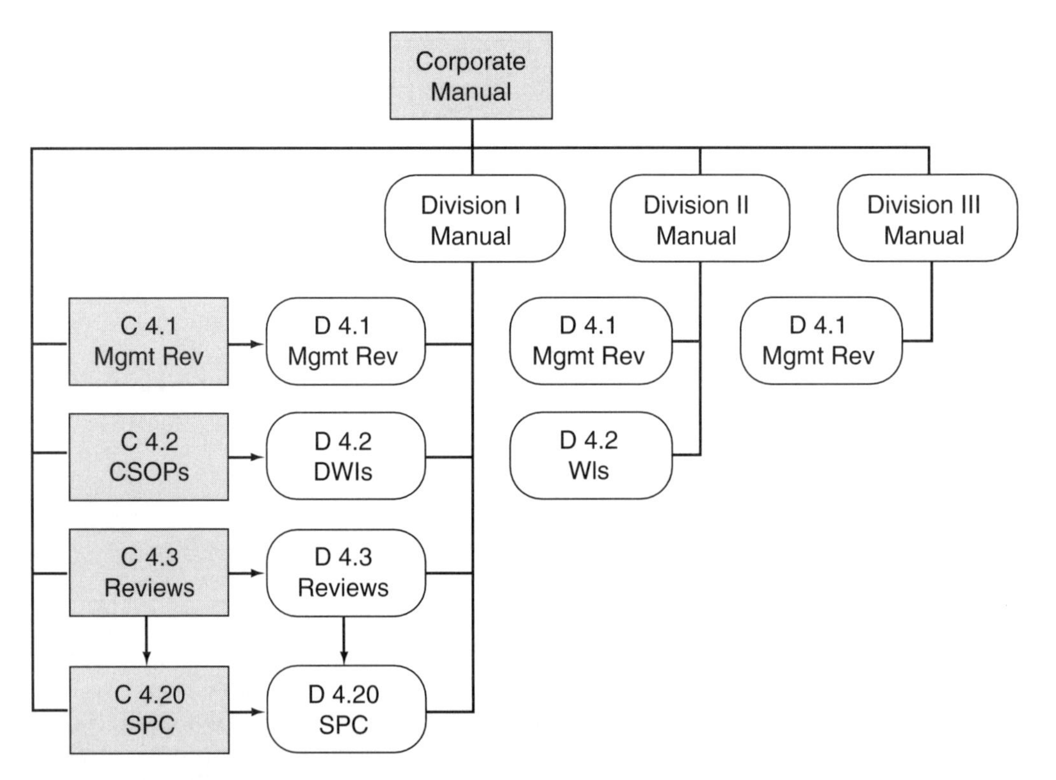

Figure 11.3 Possible configuration for a corporate/divisional integrated Manual.

Manager and was made a part of the review minutes. The minutes included the additional duties and authority of this position."

Summary

In this manner, each division responds in kind to the various divisional Quality Policy Statements in order to form a cohesive and coherent body of corporate knowledge. Each division also shares the most top-level documents, for example, Corrective and Preventive Action Procedures, Internal Quality Audit Programs, and Training Manuals. Tier III and Tier IV documents are designed expressly for use by a given division.

Conclusion

We have observed that the Manual controversy occurs because, although the supplier has chosen to write a stand-alone Manual, the writers have been inconsistent in regard to the location of the Quality Policy Statements, which often appear in the lower-tier documents.

We have found this tendency to confuse policy with process with procedure prevalent regardless of either the organization's size or industry. In many cases, the standard operating procedures (SOPs) are primarily statements of policy, and the work Instructions contain process descriptions.

We believe that a great deal of redundancy and misunderstanding could be avoided if authors always chose to use the

- Stand-alone configuration
- Hub document approach
- Manual as the location of the quality policy statements

Endnotes

1. We do not wish to imply a "dislike" of the integrated Manual. this is not the case, since in my own practice I have designed integrated Manual systems when they were required. However, as noted, this form of Manual requires considerable thought in regard to ease of use, distribution, and maintenance.

Chapter 12
Sector Specific Manuals

Part 1: The Accreditation Board Requirements

Purpose

Although it is unnecessary to be certified to ISO 9000 by an accredited registrar, accredited certifications are generally desirable due to their recognition internationally.[1] A discussion of the accreditation process is well covered in several texts, and any accredited registrar would welcome questions on this subject.[2]

Our purpose in this text is not to cover the details of accreditation but to instruct the reader in the ways that accreditation board requirements affect the Manual's structure.

Sector Specific

Accreditation boards impact a supplier's certification process via sector specific requirements that are passed on to the accredited registrar by means of memoranda backed up by a series of parallel Standards, for example,

- EN45012, General criteria for certification bodies operating quality system certification.
- EN46001, Quality systems—Medical devices—Particular requirements for the application of EN 29001.

For example, requirements of this nature strongly affect certifications in the medical industry and to some degree in software design and manufacturing.

In addition, the customer and/or the United States Government can declare sector specific conditions, for example, as in the automotive industry's QS-9000 requirements. In the case of QS-9000, GM, Ford, and Chrysler operate as a team with the accreditation boards because the QS-9000 certification is impressed upon a certified ISO 9000 system.

Direct Applicability

What we wish to demonstrate is that all of the processes and methods discussed throughout our text are directly applicable to sector specific assessments. In every case, the sequence and configuration techniques described previously will hold exactly in a sector specific Manual.

In other words the manner in which the Standard drives the lower-level documents holds true regardless of the form of the Standard. The Standard could be ISO 9001, QS-9000, FDA/CGMP, or any other. It is still necessary to respond to each requirement.

In Chapter 9, we noted that there is a flow from executive directives to management directives to operational directives. The "shalls" and "shoulds" of the Standards and guidelines define the content of the Manual. The Manual in turn confines the boundaries of the high-level procedural documents. They, in turn, drive the low-level operational documentation. This flow directives is universal.

The Audit Plan

We can demonstrate the impact of a sector specific requirement on the certification audit by means of the audit plan for elements 4.1 Management Responsibility, and 4.2 Quality System as illustrated in Table 12.1, entitled "Sector Specific Impact on ISO 9001 Audits." Notice that, although the assessor seeks answers to additional questions above and beyond the basic issues in ISO 9001, the questions are quite similar. The additional topics are highlighted in bold italics.

As indicated, more time is needed in the sector specific cases since there are more "shalls" to cover and there is an increase in concomitance; for example, there are 23 elements in QS-9000 compared to the 20 in the ISO 9001 Standard.[3]

The manner in which the supplier provides answers to the additional questions is in exactly the same way that Quality Policy Statements are used to respond to each "shall" of the ISO 9001 Standard.

We will demonstrate this technique next and take an example from each of the three specific sectors shown.

PART 2: SECTOR SPECIFIC QUALITY POLICY STATEMENTS

QS-9000

The Standard Quality System Requirements QS-9000 was developed by the Chrysler/Ford/General Motors Supplier Quality Requirements Task Force to harmonize the several quality documents already in use by those companies. The Standard is more than an interpretive guide in the assessment of automotive manufacturers. It must be adhered to in order for a company to receive a joint ISO 9001 or ISO 9002 and QS-9000 certification.

However, the degree to which this Standard is to be applied is at the discretion of the registrar and his or her accreditation board based on interpretations provided by the IASG (International Automotive Sector Group) sanctioned by the Chrysler/Ford/General Motors Task Force.

Such interpretations are published as special supplements in *Quality Systems Update,* a publication of The McGraw-Hill Companies, Fairfax, VA.

TABLE 12.1

Sector Specific Impact on ISO 9001 Audits

ISO 9001: 1994 Element	Base ISO 9001 Assessment	Sector Specific QS-9001 Assessment	Sector Specific CGMP 820 Assessment	Sector Specific ISO 9000-3 S/W Assessment
4.1 Mgmt Respon.	9:30 Scope Quality objectives Quality commit Quality org Management rep Mgmt review	9:30 Scope Quality objectives Quality commit Quality org Management rep ***Org Interfaces*** Mgmt Review ***Business Plan*** ***Analysis & Use of Company-Level Data*** ***Customer Satisfaction***	9:30 Scope ***Medical Class*** Quality objectives ***Customer Complaints*** Quality commit Quality org Management rep Mgmt review	9:30 Scope Quality policy Quality objectives Quality commit Quality org Management rep Mngt review ***Customer's Management Responsibility*** ***Org & Customer Joint Reviews***
4.2 Quality System	10:30 Quality Manual System Procedures Lower-level docs Quality planning Factored items Interface issues Currency of stnds & codes/ statutory/ regulatory	10:45 Quality Manual System Procedures Lower-level docs Quality Planning ***Control Plans*** ***Special Characteristics*** ***Use of Cross-Functional Teams*** ***Feasibility Reviews*** ***FMEAs*** ***Control Plan*** Factored items Interface issues Currency of stnds & codes/ statutory/ regulatory	10:45 Quality Manual System Procedures Lower-level docs Quality Planning ***820.20 (d)*** ***Design History File*** ***Device Master Recds*** ***Device History Recds*** ***Quality Sys Recds*** Factored items Interface issues Currency of stnds & codes/ statutory/ regulatory	10:45 Quality Manual System Procedures Lower-level docs Quality planning ***Life Cycle Planning*** Factored items Interface issues Currency of stnds & codes/ statutory/ regulator

Specific QS-9000 Shall

We will now consider a specific QS-9000 requirement and a typical response:

Section II; Sector Specific Requirements—Production Part Approval Process—General 1.1, states that

➩ "Suppliers shall fully comply with all requirements set forth in the Production Part Approval Process (PPAP) manual."

Notice that there is only one "shall," but it is a big one. In fact, the rest of the requirement gives some general direction as to the protocols for subcontracted material, questions of part need, and approval. But it is necessary to go to the PPAP itself to determine the scope of the response. This is a 51-page document published by the AIAG (Automotive Industry Action Group, Southfield, MI), and we must assume that the supplier is extremely familiar with its content.

We can now write our quality policy statements in response to the PPAP directives. To do this, we first determine where the directives belong within the Manual. This turns out to be easy since the directives can go readily into a Section 21 entitled "Sector Specific Requirements."

Our response would then be found under Section 21 of the Manual and would look something like the text below which should be an acceptable quality policy statement. In practice, a similar approach was fully accepted.

Excellent Corporation's Automotive Quality Policy Manual: Section 21—Sector Specific Requirements

Production Part Approval Process (PPAP)

Procedure

Excellent's standard practices and procedures used in the Production Part Approval Process (PPAP) are described in SOP# MNFG-2-21-001, entitled, "Production Part Approval Process."

All procedures are based directly upon the AIAG publication entitled "Production Part Approval Process." Excellent contacts its customers directly when clarification is required in regard to this directive.

Responsibility

Excellent's quality assurance manager and quality control supervisor are responsible for the coordination and completion of the PPAP activity. This activity includes completion of the required documentation and submission of the appropriate PPAP documents to the customer.

Process

At Excellent, production part approval is always required prior to the first shipment of new parts, correction of discrepancies in shipped parts, and for modified parts managed and recorded within the ECO process.

Notification of the customer by Excellent when parts are submitted for approval, unless waived by the customer, is the direct responsibility of the quality assurance manager.

Excellent customers specify the submission level they require for each part on their initial purchase order. When no submission is specified, Excellent defaults to levels specified in the PPAP.

Submissions
Documents used in a PPAP submission include the part number, change level, drawing date, and identification as an Excellent part. Measurement system variation studies are conducted in accord with customer requirements. Special characteristics are referred to as critical, key, safety, significant, major, and minor. A Ppk index is used to determine acceptable levels of preliminary process capability.

Quality
Excellent's quality control department provides dimensional analysis, material tests, and performance test analysis based on the AIAG publication, "Fundamental Statistical Process Control."

Records
The quality control department maintains all process records and engineering changes, and retains master samples.

CGMP

FDA/CGMP 21 CFR Part 820, Part VII, Quality System Regulation, Department of Health and Human Services is a mixture of ISO 9001 and the previous Part 820. It is an integral part of any accredited assessment that involves medical devices. The degree to which this Standard is applied depends on both the class of devices manufactured and the discretion of the registrar and his or her accreditation board.

Specific Shall

We will examine one of the CGMP (Current Good Manufacturing Practices) requirements and create a Quality Policy Statement in response.

> ⇨ "Sec. 820.70 Production and process controls. (d) Personnel. Each manufacturer shall establish and maintain requirements for the health, cleanliness, personal practices, and clothing of personnel if contact between such personnel and product or environment could reasonably be expected to have an adverse effect on product quality."

When we analyze the requirement we see that there is only one "shall" but there are eight directives:

- Establish requirements for the health.
- Establish requirements for the cleanliness.
- Establish requirements for the personal practices.
- Establish requirements for the clothing.
- Maintain requirements for the health.
- Maintain requirements for the cleanliness.
- Maintain requirements for the personal practices.
- Maintain requirements for the clothing.

EXCELLENT CORPORATION'S MEDICAL DEVICE QUALITY POLICY MANUAL: SECTION 9—PRODUCTION & PROCESS CONTROL

Personnel

The Excellent Corporation standards for health, cleanliness, personal practices, and use of clothing apply to all personnel in contact with either medical components or finished goods in the Clean Room.

SOP# Mnfg-2-09-011, entitled "Clean Room Dress Code and Regulations," defines the necessary procedures to ensure that these standards are implemented and maintained overall by the manufacturing manager.

Dress Code—Company-issued uniforms are worn by all personnel while working with the components and products. For safety as well as sanitary reasons, strict rules apply in regard to shoes worn in manufacturing, assembly, packaging, warehouse, and laboratory areas.

Also, strict rules apply in regard to hair and beard covers, and the use of make-up and hand cream.

The Clean Room supervisor is responsible for the strict adherence by all Clean Room employees to this standard.

Facilities

Locker, Wash, and Coat Rooms—All employees and visitors to the manufacturing and assembly areas use approved entrances and exits. Lockers are available so that street clothes and personal items can be stored before entering the manufacturing areas. Personal cleanliness is required after rest room use.

Area supervisors monitor and enforce all aspects of this Standard.

We can now write our Quality Policy Statements in response to the eight directives.

To do this, we first determine where the directives belong within the Manual. The Subpart G under which this requirement resides is entitled "Production and Process Controls," so it obviously belongs under 4.9 Process Control of the Standard.

Our response would then be found under Section 9 of the Manual and would look something like the text below. Notice that we have responded in reasonable detail to all eight directives.

ISO 9000-3

ISO 9000-3:1991 is the Guideline for the Application of ISO 9001 to the Development, Supply, and Maintenance of Software. It is to be used in general as an interpretive guide in the assessment of organizations that manufacture software. The degree to which this guideline is to be applied is at the discretion of the registrar and his or her accreditation board.

TickIT

An accredited certification called "TickIT" may be obtained in concert with an ISO 90001 initial assessment. At least one third-party assessor will need to be registered as a qualified TickIT auditor under The National Registration Scheme for TickIT Auditors.[4]

Exercise

Let us consider a specific ISO 9000-3 requirement and a typical response:

Par 5.1 Quality system—Life-cycle activities—General, states that

➪ "A software development project should be organized according to a life-cycle model. Quality-related activities should be planned and implemented with respect to the nature of the life-cycle model used."

Notice that there are two "shoulds," and one of the "shoulds" has two requirements, so our first step is to break the paragraph down into its three directives, that is

- A software development project should be organized according to a life-cycle model.
- Quality-related activities should be planned with respect to the nature of the life-cycle model used.
- Quality-related activities should be implemented with respect to the nature of the life-cycle model used.

We can now write our Quality Policy Statements in response to the directives. To do this, we first determine where the directives belong within the Manual. We can use Annex A (Informative) Cross Reference between ISO 9000-3 and ISO 9001—which is part of the guideline—to give us some direction. But it is only a general direction towards element 4.4 Design Control of the Standard.

An example of a software life cycle is defined in ISO 10005:1995 Quality Management—Guideline for quality plans.

We chose to place the requirement specifically under clause 4.4.2 Design and Development Planning of the Standard since that clause discusses time lines.

Our response would then be found under Section 4 of the Manual and would look something like the text below. Notice that we have responded in reasonable detail to all three directives.

EXCELLENT CORPORATION'S SOFTWARE QUALITY POLICY MANUAL: SECTION 4—SOFTWARE DESIGN CONTROL

Software Design and Development Planning

Procedure

Excellent's software design and development phases are documented in SOP# Eng-2-04-001, entitled "Software Design Control

continued

Guidelines."

S/W Life Cycle
All Excellent software engineering design and development projects are systematically planned by department managers and reviewed by the chief software designer before release. The computer-generated software project plan considers all aspects of the software life cycle, from project inception through to customer service and revised software releases.

Project Control
All software designers maintain project notebooks that record the life-cycle behavior of their products until production release. Customer service maintains after-sales life-cycle records.

SQA
Software quality is the responsibility of the software designers, and verification and validation testing is provided by the software quality assurance department. Test suite development is an integral part of the life-cycle plan and is the responsibility of the software services department.

Reviews
Department managers ensure a disciplined implementation of the company's software development Standards via planned design reviews during functional and final specification stages, and alpha and beta testing.

Endnotes

1. ISO 9000 astute purchasing agents, who are already on guard because it is not unusual for an ISO 9000 certified company to ship nonconforming product (the complexities of real-life shipping schedules and rapid technical changes are always with us), may tend to value less a nonaccredited registration. However, in most cases, purchasing agents tend to do what is best for their companies, and get what they need when they need it. There are no simple answers in this case.
2. See for example, Chapter 16, Registrar Accreditation, in The *ISO 9000 Handbook,* 2nd ed., by Joseph Tiratto, now published by The McGraw-Hill Companies, Fairfax, VA. Another McGraw-Hill publication, *ISO 9000 Registered Company Directory, North America,* provides a running list of accredited and unaccredited registrars on a quarterly basis.
3. The "shoulds" of the QS-9000 Quality System Requirements, August 1995, are to be treated the same as the "shalls" of ISO 9001. "should," in this case, indicates a *preferred approach*. It is not to be confused with the "notes" of ISO 9001, which are not mandatory but are used as an interpretive aid.
4. The program is sponsored by The Registration Board for Assessors, P.O. Box 712, 61 Southwark Street, London, SE1 1SB, RBA/162/94/1, 1 January 1994.

QMS STYLES

"In physics, the interpretation of experiments are *models or theories, and the realization that all models and theories are approximate is basic to modern scientific research. Thus the aphorism of Einstein, "As far as the laws of mathematics refer to reality, they are not certain; and as far as they are certain, they do not refer to reality." Physicists know that their methods of analysis and logical reasoning can never explain the whole realm of natural phenomena at once, and so they single out a certain group of phenomena and try to build a model to describe this group. In doing so, they neglect other phenomena and the model will therefore not give a complete description of the real situation."*

The Tao of Physics, *by Fritjof Capra,*
Bantam Books, NY, New York, 1984, p. 27

CHAPTER 13
READERSHIP AND FORM

PART 1: POTENTIAL MANUAL READERSHIP

We now enter into the realm in which our models and design theories come together to form documentation that relates in some incomplete way to reality. Unlike the printed circuit board, we cannot hook up probes on our employees and measure currents and voltages that indicate our success or failure to meet specification.

Our problem is that we are not only limited in our ability to fully project the dynamical behavior of our employees, as they use the documents, but we cannot even predict the impact of our written words on our customers. The written Manual is a prime example of this weakness. However, as we analyze the problem, we can offer some affective solutions.

Specifically, the Manual is the most difficult ISO 9000 System document to write. Of all the ISO 9000 documents, it must appeal to the widest set of readers. As a result, the purpose of the Manual must be carefully couched in terms of its users.

We can classify the potential readers of the Manual as follows (refer to Table 13.1, entitled "Classification of Potential Manual Readers"):

Diverse Users

As can be seen from Table 13.1, the potential readership for the Manual is extremely diverse and must comply with an ever-expanding set of user needs:

PART 2: MANUAL OBJECTIVES

From the perspective of readership, we can at least attempt to define the overall objectives of the Manual:

- To clearly describe the organization's QMS with enough detail to make it useful for a very wide range of readers
- To respond to each requirement of the International Standard so that the defined system has the potential to achieve the full benefits of continuous improvement—intrinsic within the Standard

Table 13.1
Classification of Potential Manual Readers

Potential Readers	Includes	Reader Decision Needs
Customers	▪ Executives ▪ Purchasing agents ▪ Quality assurance managers ▪ Investors ▪ Interdivisional organizations.	▪ To audit or not to audit ▪ To buy or not to buy ▪ To invest or not to invest ▪ Initially based on the scope and completeness of the Quality Policy Manual
Employees	▪ Executives, managers, engineers ▪ Supervisors, technicians ▪ Assemblers, buyers ▪ Sales personnel ▪ Internal quality auditors	▪ Is the organization really committed to quality? ▪ How can I participate? ▪ What is expected of me as a quality person? ▪ What are the quality rules of the house?
Subsuppliers	▪ Subcontractors, vendors ▪ Interdivisional organizations	▪ What level of quality is required? ▪ How will I be measured?
Third Party ISO 9000 Evaluators	▪ Assessors ▪ Registrars ▪ Accreditation boards ▪ Third-party experts	▪ Degree of compliance to "shalls" ▪ Dedication of top management ▪ Potential for effectiveness of the quality system

- To write the Manual from the customer/client's perspective and with that as the primary motivation for clarity and preciseness

ISO 10013:1995, Par 4.2.1, clearly describes additional objectives of the Manual, for example, its use in audits, training, and implementation. Objectives are defined as from a more holistic business perspective and one that ingrains the organization's personality.

New Customers

Of all the readers, it is the customer and in particular the new customer who must rely on the Manual as a primary source of organizational quality competence. Every other reader has some degree of organizational knowledge, either through a series of business transactions or through audits.

- The new customer looks at the organization with fresh eyes based on a particular technological and financial viewpoint. As a result, if the Manual is written at a level of clarity sufficient to satisfy the new customer it will certainly satisfy any other reader.

This assumption implies that there is sufficient detail to satisfy the needs of a still-diverse set of readers—for example, executives, purchasing agents, quality assurance managers, investors, and interdivisional organizations.

PART 3: WRITING FORM

The Manual, in order to satisfy such a diverse audience and to especially meet the needs of its customers, should at least[1]

Be Written in Short, Variable-in-Length, Paragraphs and Sentences

- Use one idea per paragraph.
- Realize that you will normally write about 40 percent more than is necessary.

Use Simple Declarative Sentences

- "The planning function at Excellent is the responsibility of production control."
- "Production control supplies the materials department with a daily list of raw material requirements by means of the MRP system."

Avoid Redundancy

- For example, if the Manual contains all of the Quality Policy Statements, there is no need to restate them in lower-tier documents—a common form of redundancy (refer to Section 3, Chapter 9 for a detailed discussion of Quality Policy Statements).
- Redundancy confuses the reader and forces the assessor to compare redundant sentences. Invariably, they will differ and you may receive a nonconformance based on the degree of difference.

Stress the Active Voice (Subject > Verb > Object)

Prefer

- "The President has designated the director of quality assurance as the ISO Management Representative."

Rather than

- "The ISO management representative has been designated by the president as the ISO Management Representative."

Clearly Label Section Content

It is a good idea to outline your work before you start. In that way your writing is initially structured for clarity, and the structure will then drive the correct labeling. Of course, the outline is alive and will change on you; be prepared to suffer.

Include a Useful Table of Contents (TOC)

Be sure to indicate the relationship between the TOC sections and the specific clauses of the Standard at the highest level of the Standard as possible. For example, if Section 1 of the Manual is termed Management Responsibility and deals with all the "shalls" of the Standard's 4.1 requirements, you need only reference Section 1 to the Standard's 4.1, for example

Example of a Directly Referenced Manual TOC		
Manual Section Q9001–1994 REF	**Section Title**	**ANSI/ISO/ASQC Element Number**
Section One	Management Responsibility	4.1
Section Two	Quality System Overview	4.2
.	.	.
Section Twenty	Analytical Techniques	4.20

However, if the format is such that Management Review is in Section 2 of the Manual, it will be necessary to reference that specific paragraph to paragraph 4.1.3 of the Standard, for example

Example of an Indirectly Referenced Manual TOC		
Manual Section Q9001–1994 REF	**Section Title**	**ANSI/ISO/ASQC Clause Number**
Section One	The Quality Policy	4.1.1
Section Two	Management Review	4.1.3
.	.	.
Section Twenty	Corrective Action	4.14.2

Minimize Organizational Jargon but Keep the Industry Language

Acronyms such as CEO, COO, and CFO are fairly well recognized internationally. However, short forms like DQA (director of quality assurance), and DCA (document control administrator), which may or may not be familiar within the organization (and I will tell you from experience that they seldom are), place a burden on the reader that in most cases will simply turn them off.

As big a pain in the neck as it is, it is far better to spell out the titles every time than to rely on the reader's memory. It will be appreciated. However, do not throw away the language of the industry. The Manual is read by experts, so write to their level.

Write to Be Understood, Not to Impress

Contrary to the opinion of many, "stream of consciousness" technical writing is a quick way to lose your reader. Just think about the last time you tried to read a specification sheet that used hundred-word sentences. Although they certainly are impressive, they have minimum affective value.

Clearly Define Terms

It is far better to define the terms in the text as they appear. However, a glossary is an effective back-up approach. Where there is a conflict with ISO terminology, it is practical to include a list of definitions, for example, use the word "supplier" instead of "subcontractor," or "training manual" instead of "training procedure."

Effectively Link the Reader to Referenced Documents

Once you have the reader's attention, you want to keep it. A clear reference to the associated lower-tier document maintains the interest. There are a number of ways to link, for example

- **A direct reference:** "The process for contract review is described in Doc# 3-2-001-9704, entitled, Standard Operating Procedure for Contract Review."
- **An indirect reference:** "The lower-tier documents for contract review are listed in Appendix A, entitled Master List (or Document Tree) of Lower-Tier Documents by Manual Section. The related documents are found under the column (tree) denoted as Contract Review."
- **By Hyper-link:** "Please use the ICON entitled "Design Control Process " for further information."

Use "Bullets" or Equivalents

Technical information overload occurs much faster than one would think, even among those immersed in the subject,[2]

Avoid Words That End in "-ing"

Prefer "Presidential Responsibilities:

- Assign the ISO management representative.
- Chair the management review.
- Approve the business plan."

Rather than,"The duties and responsibilities of the president, including review, are the following:

- Assigning the ISO management representative
- Chairing the management review meetings
- Approving the business plan"

Use the Spell Checker, and Then Don't Believe It

Because "manger" and "manager" are O.K. with the spell checker. So are "know" and "now," "he" and "the," and "through" and "thorough."

Use Graphics Whenever Possible, for Example, Tables, Figures, Flow Charts

This is especially true for on-line systems where only a portion of the total document can usually be seen at a time. It would be nice if all documentation systems were in the form of portrait displays on newspaper-sized screens showing multiple pages, but this is not available today.

Avoid the Future Tense; Stay with the Present

Avoid the "future" tense. Either it is happening or it is not. The use of "shall" and "will" leaves the issue hanging and cannot be readily audited.

"Shall" is used in the Standard because it is a future requirement of the supplier. Once the supplier has responded to the "shall" it is now in the present.

Prefer

- "All department managers hold monthly quality review sessions with their staff."

Rather than

- "Department managers shall hold a monthly quality review session with their staff."

Of course, if you do want to include a future event, the future tense is appropriate, for example, "In 1998, the present Tracking System will be replaced with an MRP system."

Part 4: Which Comes First? The Manual or the Procedures?

Methods

The creation of the ISO 9000 documentation system is an iterative process whereby each document tends to support other documents. As a result, the question arises as to which document to start with.

There are essentially three ways to approach this question, and all three will have some characteristics of the others in practice:

- **Top Down Method**—Begin with the Manual's quality policy statements and then create the lower-tier documents.
- **Bottom Up Method**—Work from the set of lower-tier documents to create the quality policy statements in the Manual.

TABLE 13.2
Comparison of Documentation System Methods

Method	Suggested Applications	Comments
Tops Down Manual ⇘ Lower-Tier Documents	Small organizations with minimal documentation— ▪ Metrology Laboratory ▪ Component Repair Co	▪ Tendency to stress policy over procedure. Not a great impact since OJT is prevalent.
Bottom Up Manual ⇖ Lower-Tier Documents	Large organizations with mature documentation systems based on ▪ Mil-Q-9858A or ▪ FDA/GMP Standards	▪ Little opportunity to streamline the existing system. ▪ Redundancy and obsolescence needs to be looked at in detail.
Process Flow Manual ⇧ ⇘ Process ⇗ ⇳ ⇘ Lower-Tier Documents ⇗	▪ Appropriate for all types of organizations ▪ Generally by means of a flow chart or flow table ▪ Found to be the most effective method for explicit organizational knowledge	▪ Tends to minimize total document quantity. ▪ Exposes weaknesses in operational activities. ▪ Readily catches the most frequently found gap—the handshake from dept. to dept.

- **Process Flow Method**—Start with a flow analysis—usually a flow chart—of the organization's processes and create the Manual and lower-tier documents concurrently.

Table 13.2, entitled "Comparison of Documentation System Methods," indicates which method might be the most effective for a given size organization with a given degree of documentation maturity. We have observed all of the methods in practice to some degree and believe that the most effective approach overall is to begin with an analysis of the organization's processes first, i.e., the Process Flow Method.[3]

PART 5: PROCESS FLOW CHARTS

There are a number of flow chart software programs available that are quite capable of clear process flows. The key thing to remember, however, is that a flow chart without reference to associated documents is only half the story.

In addition, not all information can be readily placed in a flow chart without obscuring its clarity. As a result, there is always room for supplemental text and complementary tables—lists of document numbers, forms to be used, and special instructions to the user.

- **Primary information**—most importantly, if a flow chart is chosen as the means of communication, it should be the primary source of information. If it shares the same data with another document, there is an excellent chance for redundancy and the tendency to confuse the reader. We have observed flow charts used successfully in both the Manual and in lower-tier documents.
- **It's the same suit!**—Just like the advertisement that stresses "that the same suits are sold in big-name shops for more money"—even though the driver keeps arguing that all the suits can't be the same if some go to big name shops and some to the low price outlet," the combination of a supplemental text and a flow chart form the informational document. *It's the same document!*

We have observed that this concept is difficult to grasp. As a result, we have described this issue in Figure 13.1, entitled "A Typical Flow Charted Procedural Structure." The details of a typical flow chart are shown in Figure 13.2, entitled "Example of a Flow-Charted Procedure." Notice the use of documentation references in both of the examples.

Imports

There is a caution on flow charts in regard to computer files for an on-line system. In some cases, the files for the supplemental text and the files for the flow charts cannot be readily integrated due to import issues. This complicates the search function somewhat but is resolvable with training.

It is best to check out the import characteristics of the software so that the ease of flow chart usage and its inherent clarity is not invalidated.

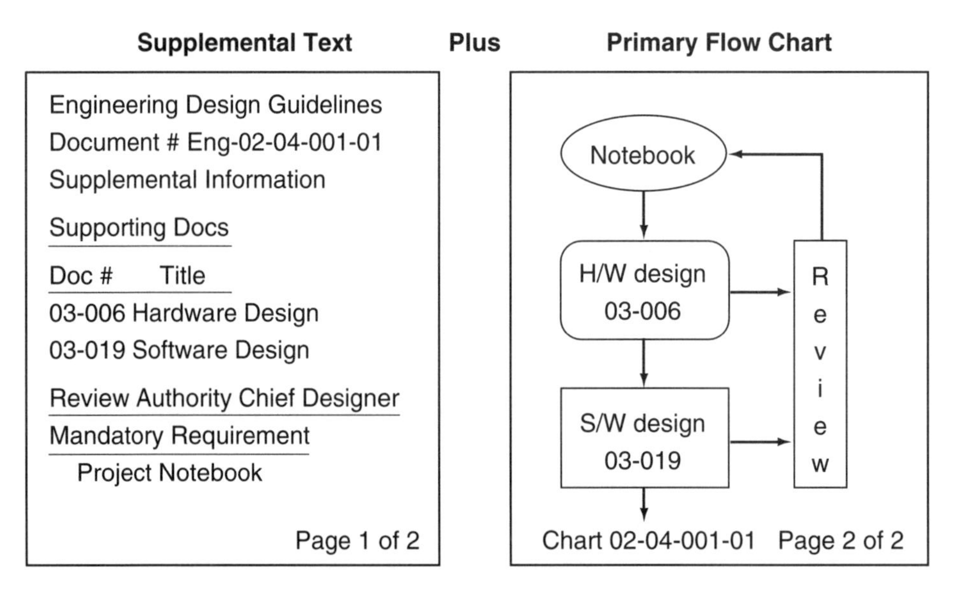

Figure 13.1 A typical flow-charted procedural structure.

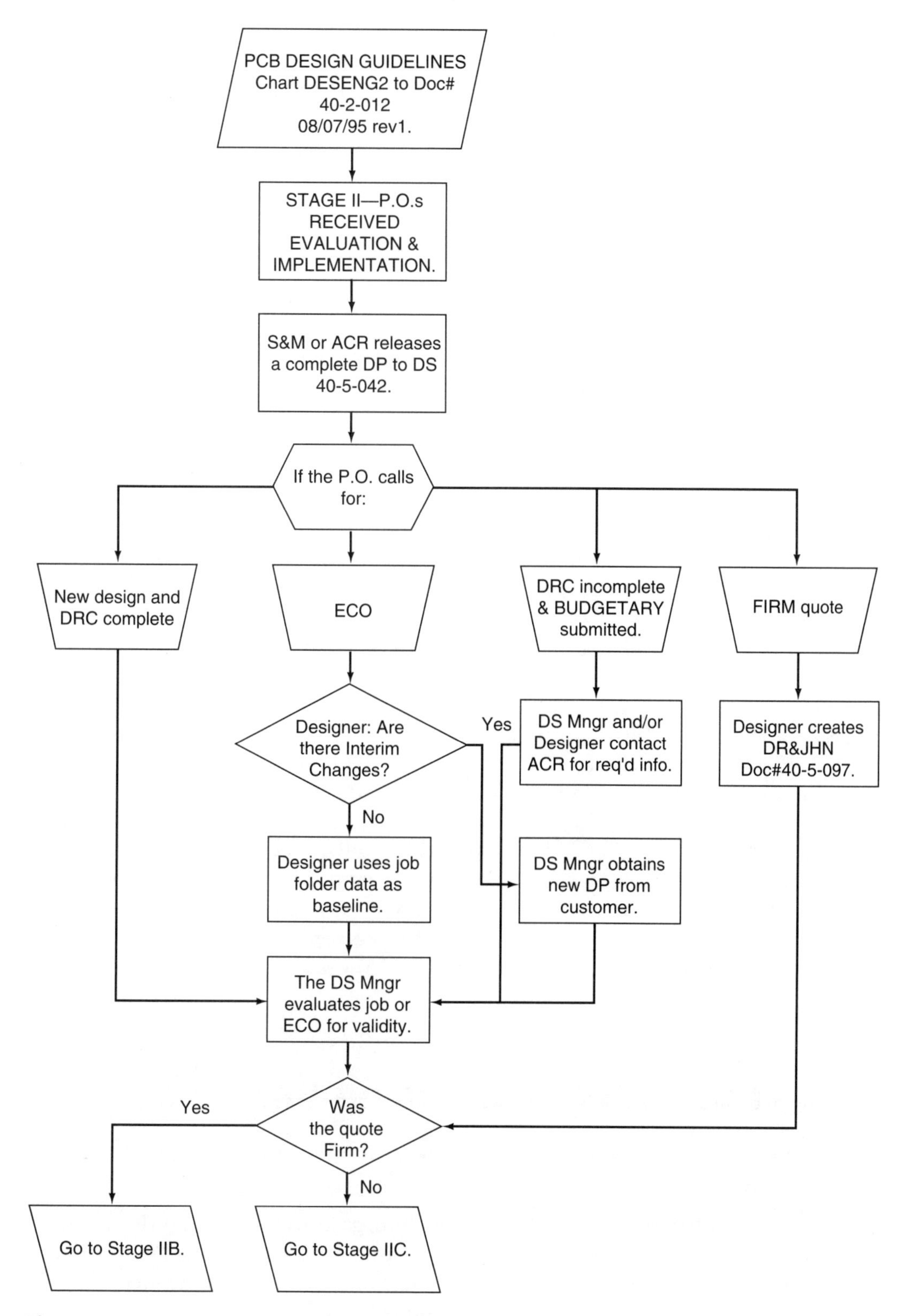

Figure 13.2 Example of a flow-charted procedure.

Part 6: The Process Document (Tier II or Hub document)

The Tier II document is expressed in many different ways, all of which are identical, for example

- Standard Operating Procedure
- Process document
- Hub document
- Quality procedure

The role that the document plays in the QMS is to describe the time-dependent behavior of the system after it has been defined as policy. Throughout our discussion we have seen several examples of this type of document, and we now wish to generalize its usage.

The importance of the process document cannot be overly emphasized. It is actually the first document that should be drafted prior to any other, including the Quality Manual (alternatively, the Quality Policy Manual).

The process document is the "homework" part of the QMS design. It is the research mechanism that establishes whether or not we really understand the organization's dynamics. Incomplete as it might be, it is the document that models the organization's ability to manage change and continuous improvement.

Although there is no specific way to write the document, we will demonstrate two methods that can be used to compose the initial draft: a "process table" approach and a cyclic "flow chart" approach.

In the first exercise, the steward for element "4.9 Process Control" has brought her team together and created a department-to-department process document that carefully maps the interfaces or "hand-offs" from one department to another. The results are shown in Figure 13.3, entitled "Excellent's Operational Processes".

In the second exercise, the Steward for element "4.1 Management Responsibility" has brought the executive group together and created a cyclic flow chart to diagram the entire development process. The results are shown in Figure 13.4, entitled "Excellent's Total Business Processes."

Part 7: Par 4.2.2 of the Standard

The Standard does imply a form of style. This is found in Par 4.2.2 Quality System Procedures, which states

⇨ "For the purpose of this International Standard the range and detail that form part of the quality system shall be dependent upon the complexity of the work, the methods used, and the skills and training needed by personnel involved in carrying out the activity."

This is a very powerful clause in that it allows the supplier to custom-fit the documentation to the organization's sophistication and complexity. Certainly, a

Phase	Description	Activities (Documentation)
One	Contract Review	▪ The VP marketing & sales supports the manufacturing process via contract review. M&S supplies the release package to production control. ▪ (S&M 001)
Two	The Build Plan	▪ The VP manufacturing prepares the build schedules and reviews with purchasing. ▪ (MNF 001, Stage 1)
Three	Purchasing	▪ The purchasing manager negotiates and procures material required to meet kit schedules. ▪ (PUR 001, Stage 1)
Four	Receipt of Raw Material	▪ The receiving & shipping department verifies and receives raw material and quality assurance runs acceptance testing. ▪ (MNF 001, Stage 2; and QA 002)
Five	Stocking Raw Material	▪ Production control stocks, releases, and inventories raw material. ▪ (MNF 001, Stage 3)
Six	Kitting	▪ The materials department kits raw material and transfers the kits to production control. ▪ (MNF 001, Stage 4)
Seven	Assembly & Test	▪ Production assembles and tests boards and integrates box level product. ▪ (MNF 001, Stage 4; and WI series 1-001 through 1-087)
Eight	Shipment to Customer	▪ The shipping department verifies customer order/pick ticket, final configuration and testing, packaging, and shipping documentation and delivers product to customer. ▪ (MNF 001, Stage 5; and W/I series 2-001 through 2-016)

Figure 13.3 Excellent's operational processes.

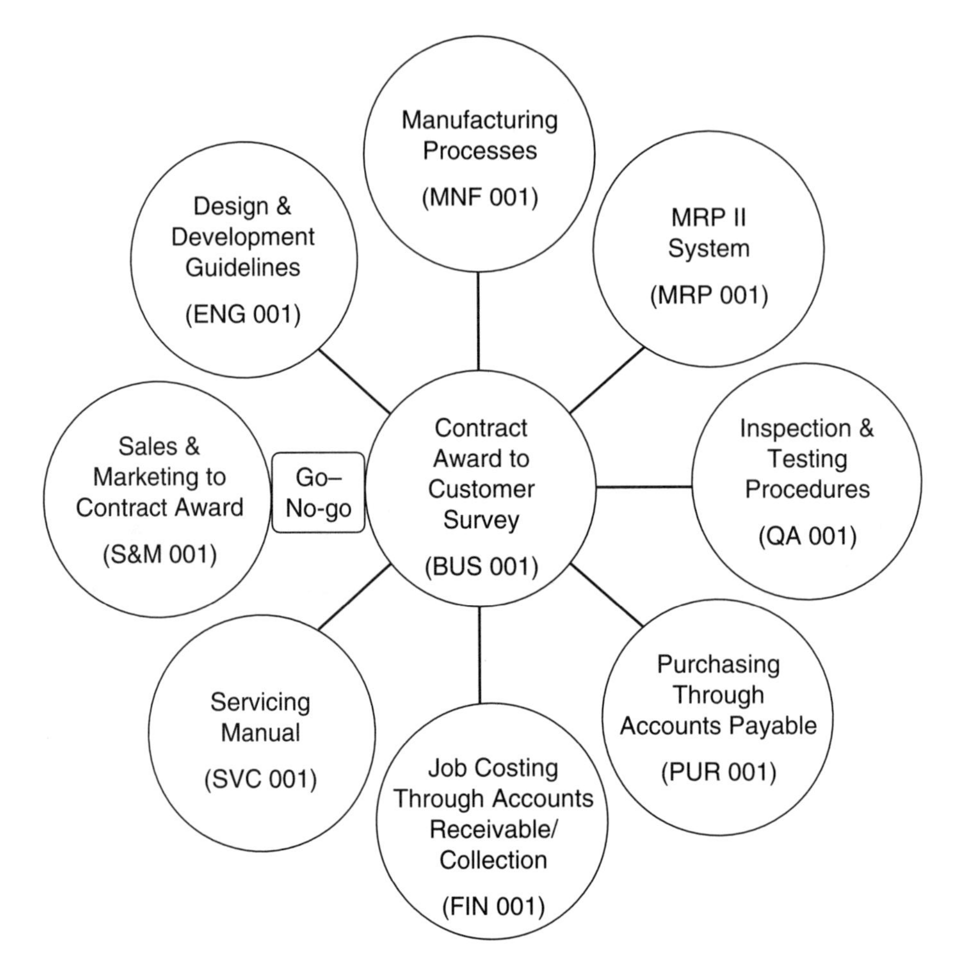

Figure 13.4 Excellent's total business processes.

design engineer requires more guideline than procedure, whereas an assembler requires more procedure than guideline.

Although it seems intuitively obvious that

- Design engineers require primarily guidelines
- Experienced machinists need a minimum of procedures
- Test operators need more detailed process sheets to properly perform their functions

It is common, for example, to find engineering documents at a level of detail that is literally overwhelming. This issue can be effectively resolved by use of a controlled one-page flow diagram that references the key design documents. This becomes a handy wall-reference chart for the project engineer.

It is also common to find Tier II and related Tier III documents that contain nearly 50 percent of the same text. However, on the first final edit it is better to be cautious so that the "baby is not thrown out with the bathwater." We maintain that

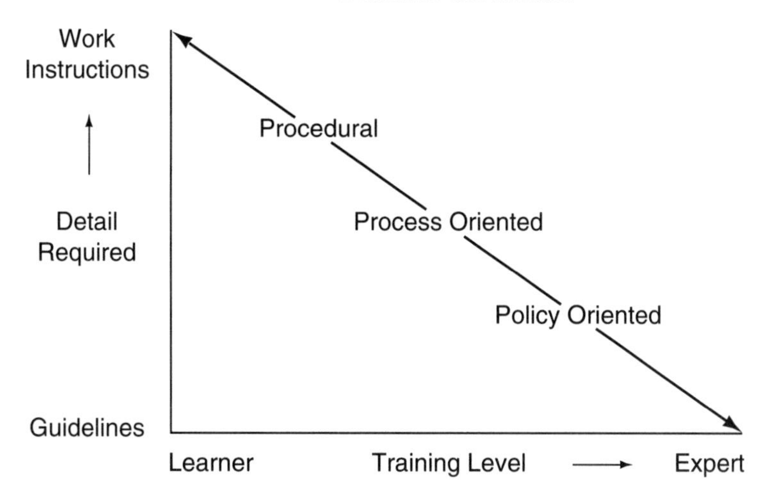

Figure 13.5 Documentation rule of thumb.

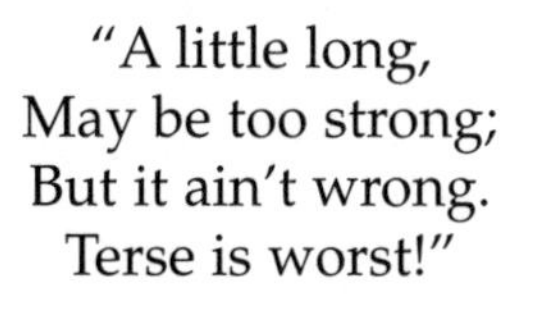
"A little long,
May be too strong;
But it ain't wrong.
Terse is worst!"

Linear Estimate

At the risk of oversimplification, we offer a rule of thumb in regard to necessary detail (refer to Figure 13.5, entitled, "Documentation Rule of Thumb"). The rule assumes a linear function, which is certainly not true, but it should clarify the point. Notice that more training is usually required as the subtlety of the documentation context expands, for example, the ability to understand work instructions versus policy statements.

Conclusion

We do not imply that everyone needs to be a great writer—we are not sure how to define such a concept. But it is within each author's ability to stress clarity and to simplify ideas.

If your text looks long and cluttered—it is. If your key ideas do not show up against the background of words—they are lost. Clear exposition is based on some basic rules, but it is also based on intuition and common sense.

The rule of thumb is one draft, and two subsequent editorial reviews with the document users before document control is initiated. Submission too early to document control is an enormous waste of time and energy; later is better. Some companies do not place their documents under control until just prior to the readiness (pre-assessment) audit.

❑ Suggested Style ❑

Our observations in over seventy organizations have indicated that, when the Manual is written for the customer, and especially for the new customer's eyes—in a clear, concise manner and filled with specific information for decision makers—it is an effective document for all other readers.

Endnotes

1. There are dozens of good books on clear writing. One Very nice little one is *Writing with Precision,* Jefferson. D. Bates, Acropolis Books LTD, Washington, DC, 2009, 1985. The old reliable one is *The Random House Handbook,* Frederick Crews, Random House, NY, 1974. The issues of manual style are also addressed in publications such as "12 Rules to Make Your ISO 9000 Documentation Simple and Easy to Use" C.W. Russ Russo, *Quality Progress,* March 1997, p. 51.
2. The seminal work of Robert E. Horn of Information Mapping, Inc., Waltham, MA, has demonstrated that the effective communication of technical concepts requires a rate of approximately three to five ideas at a time as a specific block of information.
3. The reader may feel that there is a conflict between the use of "process' analysis as the first element of system design and the belief that the first major gate in system development is the Manual. We maintain that when we begin with "process" and then use that information as a data base to define our quality policy statements, the policies end up with greater credulity and usefulness. The process documents act as the baseline research tools and the Quality Policy Manual acts as the system controller.

CHAPTER 14
THE ADVERSE EFFECTS OF PARAPHRASING

PART 1: THE TWO CLASSES OF PARAPHRASING

The Issue

We have set aside a full section to deal with paraphrasing because, in our experience, paraphrasing trivializes the Standard—although those who tend to paraphrase believe it to be a form of simplification, that is, minimalism. We intend to show that it is much more a form of nihilism (repudiation).

Classes

There are typically two classes of paraphrasing, which we define as

Class I—A direct restatement of the Standard with minor modifications.
Class II—A table of contents list of where to find information in the lower-level documents based on a direct restatement of the Standard.

PART 2: PARAPHRASED CLASS I CHARACTERISTICS

The best way to define Class I paraphrasing is to give an example of a paraphrased quality policy statement. We have chosen the ANSI/ISO/ASQC Q9001-1994 requirement, 4.17 Internal Quality Audits.

ANSI/ISO/ASQC Q9001-1994

The actual text from the Standard is

- ⇨ "The results of the audits shall be recorded (see 4.16) and brought to the attention of the personnel having responsibility in the area audited.
- ⇨ The management personnel responsible for the area shall take timely corrective action on deficiencies found during the audit."

Direct Paraphrase

A typical direct paraphrasing of this requirement is

- The results of the audits shall be recorded and brought to the attention of the personnel having responsibility in the departments audited. The department managers responsible for the area shall take timely corrective action on deficiencies found during the audit.[1]

ISO 10013:1995

In contrast, the ISO 9000 Guideline on Quality Manuals, ISO 10013:1995, gives the following example of how to respond to this requirement

- ⇨ "4.17.4.4: Reporting of results—A report is made up in conjunction with each audit, containing particulars of the object of the audit, the requirements applied as a basis and any identified nonconformities with requirements. The audit report is distributed to the manager(s) concerned. Quality system audit observations are reported in forms of the type shown in Appendix 9.—
- ⇨ 4.17.4.5: Decisions and actions—The manager of the function concerned is responsible for ensuring that decisions and actions with regard to any notified observations are taken as soon as possible—
- ⇨ 4.17.4.7: Management review of audit results—Results of audits and observations made during follow-up are presented at management reviews by the manager of the Quality Department. See Section 4.1 of this Quality Manual."[2]

Discussion on the Direct Method of Paraphrasing–Class I

In the direct method of paraphrasing, all or nearly all of the Standard's text is used as a quality policy statement. As a result, a Manual written in this fashion

- Looks and feels like the Standard itself.
- Has little to differentiate the text from that of a competitor.
- Fails to define the prescriptive "rules of the house."

A purchasing or quality assurance manager who receives such a Manual during a make–buy decision would have little information to go on. Any employee who read the Manual would be hard put to understand the dynamics of the organization and its commitment to ISO 9000.

If we compare the direct method of paraphrasing with the ISO 10013 guideline (see Table 14.1, entitled "Attribute Comparison of ISO 10013 versus Direct Paraphrasing"), we see that the contrast is significant in terms of information transfer and clarity.

- The response in ISO 10013 offers a look into the actual operation of the company, while the paraphrased text could be written about any number of competitive organizations. The competitive advantage is negated.
- The paraphrased text maintains the future tense, so it is not clear if this is what is "now" or "later."

TABLE 14.1

Attribute Comparison of ISO 10013 versus Direct Paraphrasing

Attribute	ISO 10013	Directly Paraphrased
"Shall" response	▪ Each "shall" responded to with a Quality Policy Statement.	▪ 👎 Each "shall" restated without a prescriptive response.
Clarity/tense	▪ Simple declarative sentences in the present tense.	▪ 👎 Restatement in the future tense. ▪ 👎 Questions arise as to whether or not the action has happened yet.
Detail	▪ Sufficient for decision makers to make judgments about the efficacy of the QMS and to prepare an audit plan.	▪ 👎 Nebulous information—restates the Standard.
Responsibility	▪ Clearly stated.	▪ Does imply some responsibility.
Market Differentiation	▪ Contains the personality and pulse of the supplier.	▪ 👎 Looks and sounds like everybody else.
Reference to Procedures	▪ Directly stated as Document QA 123-4.	▪ In general, also directly stated.
Configuration	▪ Directly sequenced to Standard's elements.	▪ Directly sequenced to Standard's elements

- There is nebulous information in the paraphrased text, so any decision maker would find it difficult to decide on the depth of quality in that company.

PART 3: PARAPHRASED CLASS II CHARACTERISTICS

Another technique commonly used to paraphrase is to put the language into a table of contents format. A typical pattern for 4.11.2 Control of Inspection, Measuring, and Test Equipment (IM&TE) is as follows:

Paraphrased Example

- The Excellent Corporation's IM&TE procedure demonstrates that we
- Determine the measurements to be made and accuracy required.
- Identify all IM&TE that can affect product quality.

- Define the process employed for the calibration of IM&TE.
- Identify IM&TE with suitable indicators or approved ID records.
- Maintain calibration records.
- Assess and document the validity of previous inspection and test results when IM&TE is found to be out of calibration.

Discussion of the TOC Approach to Paraphrasing—Class II

The table of contents approach at first glance almost sounds and looks like a prescriptive set of quality policy statements. If we rewrite a few of the bullets it will become easier to see the difference when responsive statements are actually used.

Nonparaphrased Response

- The identification, calibration, and adjustment of all equipment at the Excellent Corporation is the responsibility of operations engineering. Calibration plans are managed via "logs" that are maintained to indicate calibration cycles and frequency.
- Calibration labels are used on all test, measurement, and inspection equipment to alert operators that calibration is adequate or due. If calibration is overdue, operators are to immediately alert operations engineering and suspend using the equipment until calibration is completed.
- All equipment is sent out for independent calibration to companies selected on their capability in regard to using nationally known Standards. Operations engineering maintains logs of all of these transactions. A Paradox database file, CALIBRAT.DB, is maintained listing calibration status for all equipment on a calibration cycle.

Comment—Obviously, the feel is very different in the nonparaphrased response, and the information transmitted is far more useful to any reader of this Manual. The rules of the house are clearly stated with respect to inspection, measuring, and test equipment.

A quality assurance manager who was interested in gauging the metrological competency of Excellent would be favorably impressed. In actual practice, he or she is deeply impressed.

It is important to note that the previous responses require that the section be written by someone who is an expert in the IM&TE area. An experienced third-party assessor can tell within minutes how the Manual was written and whether or not it is technically sound.

Part 4: Conclusions

As demonstrated in this chapter, we believe that the use of paraphrasing is a matter of one's decision to obscure rather than clarify. The paraphrased approach

- Trivializes the Standard.
- Minimizes the opportunity to question the processes of the organization and thereby improve it.

Such trivialization is anathema to the purpose of ISO 9000, which is to support continuous improvement. In fact, in the case where one individual writes the Manual and also paraphrase's the Standard, the loss of clarity is more pronounced. Yet, it is our experience that such events frequently occur.

Therefore, although we have found paraphrasing often used, we feel that paraphrasing (restatement of the Standard) is inherently an ineffective and inappropriate method of communication.

We maintain that paraphrasing

- In its restatement of already known phrases of the Standard, offers the reader minimal organizational information and obscures the uniqueness of the quality system; it is out of phase with the concept of information technology flow.
- Is clearly not in the spirit of ISO 10013:1995 (page 11), which demonstrates that sufficient detail is necessary to paint an operational picture of the organization's response to a given requirement.
- Does not permit the reader to grasp the organization's technical and manufacturing personality.
- Negates the marketing and sales potential of the Manual and makes every organization sound like every other organization; we are befuddled as to why any organization would want to be seen as undifferentiated from its competitors.
- Is a form of intellectual dishonesty, since it trivializes the intent of the Standard, which is to provide a clearly stated and definitive top down, executive view of the organization.
- Has been found to be a tool used primarily by a single author instead of a group of technical experts, thereby diluting the technical integrity of the Manual.

Third-Party Impact

We demonstrated several approaches to paraphrasing and indicated the difficulties inherent in such an approach from a general viewpoint. We have also observed that usually, in the case of an accredited third-party assessment,

- Direct Paraphrasing is not acceptable because it cannot be clearly audited. It forces the auditor to make assumptions regarding the policies of the organization. They are not permitted to do so.
- Table of content technique is also not acceptable for the same reason since the assessor is still forced to assume what the policy (rules) are.

Alternatively

However, there are registrars and assessors who do not agree with this conclusion, and the reader who feels the same should be able to find a suitable partner for the certification effort.

- Obviously, we consider paraphrasing unacceptable in any Manual.

Endnotes

1. See for example, *Randall's Practical Guide to ISO 9000,* Internal quality Audits, page 314. For those readers who feel this is an exaggeration, our experience is that it is not.
2. This information can be found on pages 380 and 381 of the ISO 10013:1995(E), Guidelines for Developing Quality Manuals.

Chapter 15
Publication Media

Part I: Selection of a Publication Medium (Hard Copy versus Electronic)

Media Types

It is also vital to select a publication medium—hard copy, electronic (on-line), or a mixture, that best suits the organization's capabilities. The document must then be controlled in some fashion, for example

- Stamped
- Controlled versus uncontrolled checkmarks
- Different colored icons or strips or pages
- Page count noted
- Provisions made for the issuance of uncontrolled Manuals, for example, to staff, employees, customers/clients, and subcontractors

Issues

What should be the exact form of the documentation system?

The answer, of course, will depend upon the specific needs of the organization. However, we can examine a generalized scenario and then look at a few specific cases to give the reader an idea of how this type of analysis is carried out.

Once a particular alternative is chosen, it will be necessary to immediately begin to train all employees on that protocol. There is no easy answer to this question, it is a matter of experimentation to determine which system best fits our objectives.

Control Issue

We begin with a decision matrix portraying control versus documentation (refer to Figure 15.1, entitled "Decision Matrix for Documentation Systems"). We examine the possibilities for either central or local area manager (LAM) document control versus the type of document—either the Manual or lower-tier documents.

The set of protocols includes on-line, hard copy, or a mixed system.

		Documents	
		Quality Manual	Lower Tiers
Control	Central	On-line Hard copy Mixed	On-line Hard copy Mixed
	LAM	N/A	On-line Hard copy Mixed

Figure 15.1 Decision matrix for documentation systems.

A hard-copy system has another degree of complexity in that the documents can be distributed as either

- A set in binders
- Individual documents

Alternatives

To limit the number of choices, we assume that our organization is characterized by

- Little or no central control, a central document control center does not exist nor does it readily fit into the economic viability of the organization
- Local area managers, department heads, who have agreed to maintain document control procedures that include locally controlled master documentation lists
- A networked system that is readily available so that Tiers I, II, and III can go on-line immediately.

Therefore, of all the many possible configurations, it appears that we have only two practical choices:

Alternative I—Keep Tier I and II centrally controlled, and have Tiers III and IV controlled by LAMs.

- Pro—the easiest to find high level documents
- Con—requires a dedicated central document control manager

Alternative II—Keep Tier I central, and have Tiers II, III, and IV controlled by LAMs.

- Pro—requires minimal central document control management
- Con—difficult to find high-level documents

Decision

In an actual case, Alternative I was chosen. The difficulty in finding high-level documents turned out to be the decisive factor. The central document control function was shared by several employees and consumed a modest amount of time once the system was in maintenance.

Part 2: Hard Copy System Issues

Control

In practice, a purely hard copy system is the most expensive and time consuming to maintain, and it is best to limit the number of controlled Manuals to essential personnel, for example, copies to the document owner (often the site manager), the ISO 9000 management representative, and the registrar.

Uncontrolled

Uncontrolled copies usually need to be released by the owner on a filtered basis, and the Manual should have some sort of disclaimer, for example, "The contents of this uncontrolled Manual may not be at the latest revision level."

Changes

It is important to minimize the number of times per year that changes are made to the Manual to control printing costs. Such costs can be very significant when you consider the expense of labor and distribution control.

It is best to collect minor changes and do a rewrite periodically unless there has been some major action taken, for example, reorganization, third-party audit that resulted in nonconformances, merger/acquisition activities, or a business scope upgrade.

Distribution

The creation and distribution of the Manual must comply with the requirements of element 4.5, Document and Data Control, of the Standard. A convenient checklist to be used to ensure this compliance is shown in Appendix E, entitled "Checklist for ANSI/ISO/ASQC Q9001-1994 Element 4.5 Document & Data Control Quality Manual Requirements."

Part 3: On-Line System Issues

Impact

With the advent of enhanced information technology networks, many organizations are either already networked or plan to be in the near future. Any move to place the Manual on-line will have an immediate impact in the ease of control. The amount of software available for on-line use is overwhelming. The platforms are either self-developed or based on readily available software.[1]

Hard Copy

However, since the Manual serves as an excellent marketing tool, we will still want to produce uncontrolled hard copies under the same conditions mentioned above. In other words, an on-line Manual tends always to end up a mix of electronic and hard copy media. This is often true for the entire documentation system, since we find that drawings, blueprints, schematics, data sheets, and production tags, for example, tend to remain hard copy.

Key Factors

The decision to go on-line involves the solution of a number of critical factors, several of which are beyond the scope of this text. However, a few key factors are

Structured Hypertext[2]—The use of hypertext alone will not guarantee an effective system unless the entire documentation structure is logically designed on the basis of hierarchal need. The old adage, "garbage in—garbage out" still holds true.

Available Expertise—Even if the choice is made to go with off-the-shelf quality management system software (QMS S/W), we have found it necessary to have someone on board who is a computer expert, in conjunction with a dynamic training program.

Most importantly, there is a clearly defined need to have support available 24 hours a day, 7 days a week. The reason for this is that QMS S/W packages are designed to manipulate *ideas* as opposed to MRP and SPC type packages which are designed to manipulate *data.* As a result, there is a constant need for clarification as to what the information means when ideas are involved. Also, unless you have personally designed the QMS S/W data bases you will be ill equipped to correct logical software glitches.

Graphics (flow charts, tables)—Although graphics—and in particular flow charts—can greatly enhance a document's overall usefulness, unless there is clear evidence that the flow charts can be effectively integrated into the document's application software, it may be better to use tables as a means of clarity. It is always best to know the limits to interoperability for your software before you invest a great deal of time and funds into any type of graphics.

Training Issues—The moment the decision is made to go on-line, the training must begin immediately. It has been our experience that on-line systems require far more training than hard-copy systems.

Projection Systems—To avoid an unacceptable level of dropped hard copy in an on-line system, for example, and in meetings, and training sessions, it is advisable to install projection systems that are driven by your computers. The issue is one of projection intensity, and it needs to be checked out before installation in order to keep everyone in the room from dozing off in front of the president.

Registrars—It is somewhat early in the on-line approach, so you will find a wide range of methods used by third-party assessors to accept and recommend certification for an on-line system, particularly when document reviews are often performed off-site. It is best to check in with your registrar and develop a mutually agreeable audit plan that will resolve this issue.

PART 4: GENERIC NUMBERING SYSTEM

In many cases, the documentation system is a mixture of on-line and hard copy. This raises the issue of just what type of numbering system will work concurrently. One possible approach is to have the on-line format as follows:

On-Line Format

Dept	Tier II	2	05	126	01
(Directory)	(Sub-directory)	(Level)	(Element of Stnd)	(Doc)	(Revision level)

- Only 8 integers or alphas are needed.
- Permits 999 Tier III documents for every Tier II document.
- The department and Tier II levels do not require a number since they are in computer directories and subdirectories.
- The level number is either 1, 2, 3, or 4 for the various tiers.
- The element of the Standard would run from 01 to 20.
- The document number runs from 001 through 999.
- Revision control runs up to 99.

Hard Copy

When hard copy documents are required, we would simply add the operating department's code name, for example, marketing & sales, and form:

MKT20300103 = Rev three of the M&S Tier II contract review document, no. 001.

Endnotes

1. The ASQ publication *Quality Progress,* issues a yearly summary of available software. See, for example, *1996 Software Directory,* April 1996. Additional publications include, "Quality Systems, Intranets: Using Web Technologies in a Regulated Environment," Morteza Minaee, *Medical Device & Diagnostic Industry,* November 1996, p. 46. Documentation on Tap, Colin Maunder, IEEE Spectrum, September 1994, p. 52; Document Shock! Stemming the Rising Document Tide," Barry Quiat, *Network Computing,* January 15, 1994, p. 82.
2. An excellent discussion on structured hypertext is given in *Mapping Hypertext,* by Robert E. Horn, A Publication of the Lexington Institute, Information Mapping, Inc., Waltham, MA, 1989.

Chapter 16
Quality Policy Manual Scope of Effort

Part 1: Estimates

A considerable effort is required by top management to produce a stand-alone, ISO 9001 sequenced, Quality Policy Manual that integrates business strategy with quality management. It is an iterative activity that peaks approximately one-third of the way into the process and then requires some level of maintenance up to the certification assessment.

After certification, maintenance is normally required prior to a surveillance assessment and when the organizational and operational structure makes significant changes.

Labor Estimate

We can estimate to some degree the number of hours required to create a fully compliant ANSI/ISO/ASQC Q9001-1994 Manual if we assume

- a process manufacturing facility
- a staff of 100 employees—20 percent of which are managers and line supervisors
- a quality assurance department
- a management representative who is also a full-time manager
- full time clerical support (at least one person)
- a part time consultant (approximately 25 percent of the time on-site during the precertification effort)
- a training program that includes documentation writing skills for some employees
- a documentation system that already exists in the form of some basic work instructions and operational formats
- a plan showing that the designated employees write, edit, and research for three hours for every hour that the consultant has been on-site.

Part 2: Discussion

The estimate scales with size and product complexity, so plus 50 percent and minus 25 percent are possible. Table 16.1, entitled "Excellent Corporation's Quality Policy Manual Timeline," illustrates a typical scenario and plan for the Manual. The time to the certification assessment is twelve months from the program kick-off date.

As indicated in Table 16.1, to create a Manual of approximately 50 pages requires a considerable effort by the entire staff—approximately 56 employee days. This is not a one-two-three exercise, and the effort includes team meetings and considerable dialogue.

As indicated, the load is greatest on QA because we have assumed that at least internal quality audits and metrology have been assigned to that group, along with inspection and testing. The potential loading on each department will become clearer as we proceed through the rest of the text.

We believe that the result of such an effort is a Manual that makes sense to all of its readers and propagates a favorable impression of the organization from both a strategic and a technical standpoint.

TABLE 16.1

Excellent Corporation's Quality Policy Manual Timeline

Manual Phases	**Scheduled Months for Actions in Grey**											
Months from Kick-off	**1**	**2**	**3**	**4**	**5**	**6**	**7**	**8**	**9**	**10**	**11**	**12 Cert**
Initial drafts due												
First draft review												
Final draft review												
First master published												
Master review after continuous improvement audit												
Master review after readiness assessment												
Master review after certification audit												
Total writer/editor/ research days												
ISO mgmt rep	32	8	8	4		2		2		2	8	2
Technical writer		40	8	4		2		2		2	4	2
Clerical		40	16	8		4		4		2	8	2
ISO administration sub-totals (hrs)	***32***	***88***	***32***	***16***		***8***		***8***		***6***	***20***	***6***
General manager	8	4	2	1		1		1		1	1	1
Engineering mngr.	12	6	3	2		1		1		1	1	1
Operations mngr	12	8	4	2		1		1		1	1	1
Purchasing mngr.	8	4	2	1		1		1		1	1	1
QA manager	16	12	8	6		4		2		1	4	2
M&S manager	8	4	2	1		1		1		1	1	1
HR manager	8	4	2	1		1		1		1	1	1
Finance manager	4	2	1	1		1		1		1	1	1
Supervisors			16	4		1		1		1	1	1
GM & Staff Subtotals (hrs)	76	44	40	19		12		10		9	12	10
Grand total (hrs)	108	132	72	35		20		18		15	32	16

Grand total of hrs = 448 employee hrs or approximately 56 employee days
ISO Admin = 216 hrs approx. 27 days GM & Staff = 232 hrs approx. 29 days

QMS DESIGN SUMMARY

"All at once it became vividly clear to Adam. He turned to the sated Eve—she was surrounded by apple cores—and wiped the apple juice from his chin with the back of his naked hand and remarked, "You know my dear, we are living in a time of transition!"

Anonymous

Chapter 17
Issue Resolution

Indeed, we do live in a time of transition as we observe the characteristic of our quality practitioners inexorably decline into mediocrity.[1] However, we believe that this is a symptom of a young industry in its attempt to mature. It will swing around, just as all things in life follow a cycle of growth and decay. We, of course, hope that our attempt to define an effective and affective set of QMS design rules will help in some measure to improve this situation.

Specifically, we have demonstrated that the root causes of the observed QMS deficiencies exist primarily in the tendency for authors to

- Lack clarity in their overall QMS structural design
- Perform inadequate research into the reality of their systems' performance
- Spend too little time in process-document creation
- Place Quality Policy Statements in lower-level documents instead of in the chosen stand-alone Manual
- Paraphrase the Standard and leave out key prescriptive details for decision makers
- By-pass "shalls" because of an incomplete analysis of the requirements
- Use an integrated policy and procedure manual that does not fully respond to the Standard's requirements ("shalls")
- Not stress the importance of Tier-to-Tier linkage

This tendency is a result of an industry-wide disagreement by ISO 9000 practitioners on the purpose and structure of the Quality Manual. The result is confusion over what constitutes an effectively written document and what specific textual tools are available to create the Manual.

Such practices are counterproductive, since they invariably produce redundancy, omission, and noncompliance with the International Standard.

Proposal

To ameliorate this situation, we have attempted to place QMS design on a scientific foundation. We have proposed a number of design techniques that we believe produce compliant Quality Manuals—and, as a result, compliant systems. Such systems integrate business strategy with quality management and thereby form the organization's total quality strategy.

A summary of the specifics of such techniques is addressed in the next chapter. In this chapter, we wish to examine the anticipated benefits of a QMS which uses the proposed tools.

Benefits

The benefits to be gained from a QMS that is fully compliant with the International Standard and integrates business strategy with quality management are summarized in Table 17.1, entitled "Benefits of the Unified QMS." We have assumed that the fully responsive techniques discussed in this treatise have been chosen to create the QMS.

Table 17.1 considers three categories of concern:

- **Type I**—Readers
- **Type II**—Organizational objectives
- **Type III**— the total quality management system

TABLE 17.1

Benefits of the Unified QMS

(fully compliant with the International Standard and integrates business strategy with quality management)

Type of Reader or Function **Type I—Readers**	**Benefits**
▪ Site Manager	▪ Significantly improved communication at all levels; opportunity to modify processes based on a more complete perspective.
▪ Executive Staff	▪ Obviously strong correlation between the completeness of the Manual and the overall knowledge of the Executive Staff in regard to business policy.
▪ Customer	▪ Dramatic improvement in communication and acceptance for more demanding contracts.
▪ Third-Party Registrars and Assessors	▪ Exceptional clarity leads to a far more effective assessment at a greater depth into the organization.
▪ Subsuppliers	▪ Significantly improved grasp of your objectives and how to respond to them.
▪ All Decision Makers	▪ The availability of clear and concise information significantly improves the decision-making process.

TABLE 17.1
Continued
(fully compliant with the International Standard and integrates business strategy with quality management)

Type of Reader or Function Type II—Organizational Objectives	Benefits
▪ Response to Organizational Objectives	▪ Exceptional response at all levels of the organization in the measurement and publication of business metrics.

Type of Reader or Function Type III—The Total Quality Management System	Benefits
▪ Tier II Documentation	▪ Very strong influence on the completeness and effectiveness of Hub documents and knowledge of business processes.
▪ Tier III and IV Documentation	▪ Appears to have a minor effect. We have observed exceptional Tier III performance with incomplete stand-alone policy Manuals.

Type I

For the readership, the benefits extend from improved communication to improved strategic and tactical decisions. Most importantly, all members of the value chain are included in this group.

Type II

In regard to organizational strategy, the impact is exceptional when we analyze the organization's ability to set, pursue, and communicate quality objectives. This enhanced communication begins with the posted "Organization's Quality Policy" and propagates throughout the organization via the publication of progress reports on measured performance.

Type III

With respect to the total quality management system, we observe a strong impact on the clarity and completeness of Tier II documents.

However, the impact on Tier III documents is not that strong. This comes as no surprise, since every organization—no matter how new or small—must work from some basic documentation and format. It is not unusual to find excellent work instructions and significantly incomplete stand-alone policy Manuals in the same organization.

Endnotes

1. Unfortunately, justifiable complaints against the ISO 9000 schema can be found throughout ISO 9000 publications, for example, *Quality Systems Update,* The McGraw-Hill Companies, September 1997, Letter to the Editor, and *Quality Progress,* ASQ Publication, September 1997, Letters, p. 6.

Chapter 18
QMS Proposal Specifics

We have now treated each element of QMS design in some detail, and are in a position to summarize the set of attributes (design rules)—which, if applied by authors to the QMS, should resolve all of the issues under consideration (refer to Table 18.1, entitled "Information Attributes Proposed for the QMS").

Summary

This completes our ISO 9000 Quality Management System Design treatment. We believe that the use of the proposed design rules will tend to resolve the Quality Manual controversy and result in a more user-friendly, effective, and affective QMS.

We have approached the subject of QMS design as a scientific exercise—although there is a very significant subjective aspect to the paradigm. In this sense, we are reminded of the words of Thomas S. Kuhn:[1]

> Probably the single most prevalent claim advanced by the proponents of a new paradigm is that they can solve the problems that have led the old one to a crisis. When it can legitimately be made, this claim is often the most effective one possible.

We trust that we have presented a legitimate discourse. In fact, it is now time to apply the QMS design rules to the ambitious and fast-growing Growth Division of the Stable Corporation. We shall see how effective our proposal will be.

Endnotes

1. Thomas S. Kuhn, *The Structure of Scientific Revolutions,* 2nd ed., Enlarged, Vol. II, No. 2, Foundations of the Unity of Science, University of Chicago, 1970, p. 153.

TABLE 18.1

Information Attributes Proposed for the QMS

Information Attributes	Benefits of Approach
▪ Integrates business strategy with quality management— ▪ by means of a total business & quality policy format.	▪ The basis of a well informed organization. ▪ Supports the organization's information technology imperatives. ▪ Caters to decision making.
▪ Complies exactly with the contractual Standards. ▪ Responds positively with a quality policy statement to each "shall," and adheres to the spirit of the ISO 9000 guidelines.	▪ Ensures compliance with the contractual Standards. ▪ Enhances the inherent continuous-improvement cycle. ▪ Enhances the possibility of payback.
▪ Utilizes stewardship management with cross-functional teams. ▪ Uses experts to write sections.	▪ Ensures that top management is committed to the complete documentation, implementation, and demonstration of effectiveness of the program. ▪ Partially satisfies the affective part of QMS design.
▪ Analyzes all processes of the organization prior to the specific design decision. ▪ Uses flow-charting methods if possible, otherwise table formats.	▪ Greatly enhances the development of the Quality Manual and provides an outstanding base for the Hub documents.
▪ Formally declare the specific sequence pattern for the QMS layout, e.g., direct-sequence with ISO 9001 elements, and intensively trains team members in the approach. ▪ Selects from: Direct ISO 9001, Shewhart Cycle, Operational Cycle methods.	▪ Team members will work toward clear linkage to lower level documents. ▪ Team members will concentrate on operational flow and the continuous improvement cycle.
▪ Formally declares the specific Quality Manual configuration, e.g., stand-alone, and intensively trains team members in the approach. ▪ Selects from either stand-alone or integrated configurations. ▪ Is consistent with the placement of quality policy statements.	▪ Team members will work toward appropriate departmental training programs; and marketing and sales will ensure the appropriate means of distribution.

Table 18.1
Continued

Information Attributes	Benefits of Approach
▪ Evolve on-line in the shortest possible time. ▪ Puts Level I & II documents on-line as early as possible.	▪ Significant gains in document control and revision control. ▪ Lowers distribution and maintenance costs.
▪ Appeal directly to the customer/client's perspective.	▪ Ensures clarity for all readers. ▪ Caters to the new customer.
▪ Clearly places all Quality Policy Statements in the Quality Policy Manual if stand-alone method is used. ▪ Avoids redundant statements in lower-level documents. ▪ States the Quality Policy Statements once and places them at the front of the written procedural sections, if integrated method is used.	▪ Creates a fully compliant Manual that is clear and precise in regard to the organization's rules, methods, and business strategies.
▪ Includes sufficient detail in the quality policy statements ▪ Allows the reader to understand how the organization actually works.	▪ Provides all readers, especially decision makers, with worthwhile pertinent organizational information. ▪ Is a highly effective document for the new customer.
▪ Provide user-friendly navigation tools. ▪ Has a four-tier structure. ▪ Utilizes Hub documents.	▪ Significantly increases the probability of effective implementation by all employees. ▪ Tends to minimize the number of documents. ▪ Clarifies linkage.
▪ Avoids paraphrasing.	▪ Removes the trivialization of the Standard, which is anathema to the quest for continuous improvement—where completeness and clarity are mandatory.
▪ Uses Effective Styles, for example: ▪ Features simple declarative sentences in the active voice & present tense, ▪ Avoids redundancy, has clear labels. ▪ Has a useful TOC, minimizes jargon. ▪ Stresss clarity, define terms. ▪ Effectively links, avoids "ings." ▪ Uses bullets; doesn't rely on the spell checker, uses graphics.	▪ Enhances information flow. ▪ Increases rate of understanding. ▪ Increases training retention time. ▪ Increases rate to find documents.

Table 18.1
Continued

Information Attributes	Benefits of Approach
▪ Uses the same attributes for sector specific requirements.	▪ The same attributes apply.
▪ Uses as many words and charts as needed to produce the organizational image that is desired.	▪ A little long, May be too strong, But it ain't wrong. Terse is worst!
▪ Offers utility—documents that ▪ Are worth reading. ▪ Contain industry-familiar phrases. ▪ Are relatively easy to obtain.	▪ Enhanced communication. ▪ A common training vocabulary. ▪ The quality policy directive propagated to all employees.

Case Study—The Growth Division

"The secret of the master warrior is knowing when to fight, just as the secret of the artist is knowing when to perform. Knowledge of technical matters and methods is fundamental, but not sufficient to guarantee success; in any art of science of performance and action, direct perception of the potential of the moment is crucial to execution of a master stroke."

The Lost Art of War, *by Sun Tzu II, commentary by Thomas Cleary, Harper San Francisco,1996, p. 52.*

"Whatever I find myself doing, I become aware that I must make a choice. I must make a choice or find the choice made for me. I must choose from whatever alternatives my experiences have stored up and from whatever alternatives my emotions made available to me. I must try to calculate the risks involved, and manage my fears while calculating."

Choice Points, *by John C. Glidewell*
The MIT Press, Cambridge, MA, 1972, p. 3

Choice Point

The Memo

The memo from Judy G., president of the Stable Corporation of Dallas, TX, to Fran L., VP & general manager of the Growth Division in St. Louis, MO, was sharp and clear:

> Fran,
>
> Your 1998 budget is admirable, but a stretch. 1997 was a grinder for you and the decline in gross margin could have been a lot worst—but you have a fine group of people and they came through in the final few weeks of the fiscal year. A little too close for comfort, but we made it!
>
> I am concerned that you may be pushing them too hard to make your numbers. Based on your fine work overall, I still want to elevate your division to a wholly owned subsidiary by 2000, so please take another look at where you are and give me some strong points on why we should accept your budget in the face of declining gross margins. I think you might spell out more clearly your cost-reduction program.
>
> David and Mark are prepared to help you—even to come down for a few days to put the last touches to the story. Don't be afraid to ask.
>
> Best regards to you and your staff,
>
> Judy.

As always, in spite of the distance between Dallas and St. Louis, Fran had kept an excellent relationship with corporate over the years, and she knew that Judy would do all she could with the board to support Growth's latest budget.

After all, the boss had come up through the ranks just as Fran had, and knew very well how tough it was for anybody to stay in top management. They had both survived and grown intellectually and technologically.

"I mean, good heavens, the boss is an EE with twenty years in hi-tech program management and an MBA, " she mused. "And I'm an IE with fifteen years in hi-tech operations, so if we don't know what we're doing by now there's no hope."

But Fran had to move fast—not just to convince the boss and the board, but to convince herself and the staff that they were not headed for disaster.

The Data

Fran's first move was to look again at the 1998 plan published by Greg C., controller of Growth. His work was always meticulous and insightful.

Fran took a pencil to the reams of data and examined the general behavior of the company in terms of sales, cost of goods sold, gross margin, engineering expenses, marketing and sales expenses, and finance & administration expenses. It was the profit before income tax that she wanted to focus on. She felt that, with the prime at around 8.5 percent a return in the vicinity of 13 percent would make everybody happy. The boss was looking at 10 to 15 percent, so Fran was in the ballpark.

What she saw raised many questions. The results are shown in Case Study Table 1—Growth Division's Financial Analysis.

It was clear that the first quarter was heading in the wrong direction, especially since the division had been getting off to a fairly strong start over the last three years. Competition for the new pipe-lined boards was fierce, and there were a lot of buy-ins that helped to drag down sales, and with it gross margin as Growth staffed to 105 employees to meet the CY 98 production goals.

CASE STUDY TABLE 1

Growth Division's Financial Analysis

Financial Analysis for Growth, Inc.	CY 94	CY 95	CY 96	CY 97	CY 98 Goal	Q1 98 Estimated
Revenue (sales) in millions $	1.10	2.30	3.85	7.90	15.50	3.10
Cost of goods sold (COGs) $	0.77	1.38	1.93	4.74	8.53	1.92
Gross margin (GM) $	0.33	0.92	1.93	3.16	6.98	1.18
GM%	30%	40%	50%	40%	45%	38%
Engineering expenses (Eng'g) $	0.17	0.35	0.58	1.19	2.33	0.47
Marketing & sales exps (M&S) $	0.13	0.28	0.46	0.95	1.86	0.37
Finance & administration (F&A) $	0.06	0.12	0.19	0.40	0.78	0.16
Profit before income tax (PBIT) $	–0.02	0.18	0.69	0.63	2.02	0.19
PBIT %	–2%	8%	18%	8%	13%	6%

Fran took the revised organizational chart out of her desk and perused the various departments for any indication of overstaffing. Based on her ISO 9000 research, the organizational chart indicated the functional relationships of all personnel in Growth along with the indicators that summarized which employees had the responsibility and authority to manage, perform, and verify work affecting quality.

Fran had attached an indicator legend to the chart for ease of use. (Refer to Case Study Figures 1, and 2, entitled "The Growth Division Organization and Responsibility" and "Responsibility & Authority Legend for Growth," respectively.) In addition, the series of numbers from 4.1 through 4.20 related directly to the Standard and indicated a potential set of ISO 9000 stewardships where each steward would be responsible for the effective documentation, implementation, and demonstration of effectiveness for his or her element. This technique would come in handy when and if the decision to pursue ISO 9000 certification was made.

Productivity was an issue and cost cutting was critical.

It was time to make a series of critical decisions with the staff at today's management review. All four directors and Greg C. were in town and had already been instructed to prepare ideas on how Growth could meet the 13 percent PBIT. In addition, Fran circulated a number of recent articles on ISO 9000 to see if they could be of any interest to the group. Several of her competitors were now certified to ISO 9001, and that was indeed a concern. The last four top management reviews had discussed this issue in some detail.

The Staff Meets

The top management staff meeting began at 3:00 P.M. and did not end until after 11:00 P.M. By 9:00 P.M., everyone was loaded with pizza and coke and donuts and coffee, and slurred speech was common.

The group had always been close in thought, especially since they had all helped to solidify the young division. Janis K., director of sales & marketing, was the newest member but she had been on-board since the fall of 1995. Greg C. had joined the operation several months prior to Janis. They were a good-natured bunch who partied well at the annual picnics and often met after work for a beer or two, but this discussion was pushing them a bit.

Janis K. had offered a number of creative approaches to increasing market share, based on Marlin P.'s optimistic new product design ideas for the multiprocessor boxes and systems. Greg's financial analysis had supported Janis and Marlin, based on prior returns from new products which were very close to the desired 40 percent of annual revenues. Karen S. was convinced that she could shave several points off the COGs through improved buyer techniques, and Mike H. felt that a stronger program in cost of quality should also pick up a few more points in GM.

The problem was that there was no clear platform upon which to launch, consolidate, and control all of the good ideas on the table.

At about 10:00 P.M., Fran decided to roll the dice on ISO 9000 and see what the reaction would be. She had researched the subject intensely and was intrigued by so many reports of payback in the form of new contracts from key OEMs and the solidarity it created within the organization.

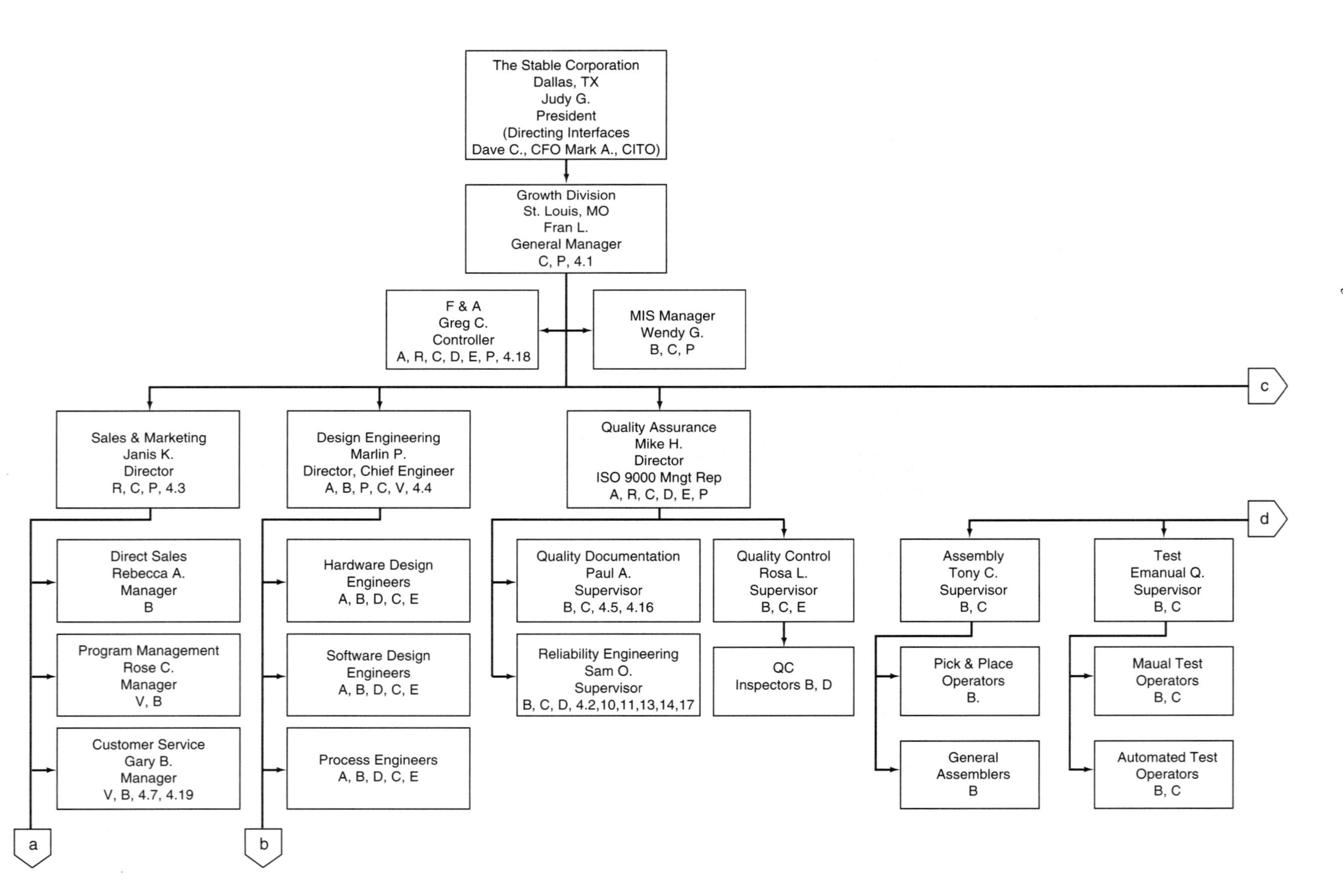

The Stable Corporation
Dallas, TX
Judy G.
President
(Directing Interfaces
Dave C., CFO Mark A., CITO)
Growth Division
St. Louis, MO
Fran L.
General Manager
C, P, 4.1
F & A
Greg C.
Controller
A, R, C, D, E, P, 4.18
MIS Manager
Wendy G.
B, C, P
c
Sales & Marketing
Janis K.
Director
R, C, P, 4.3
Design Engineering
Marlin P.
Director, Chief Engineer
A, B, P, C, V, 4.4
Quality Assurance
Mike H.
Director
ISO 9000 Mngt Rep
A, R, C, D, E, P
d
Direct Sales
Rebecca A.
Manager
B
Program Management
Rose C.
Manager
V, B
Customer Service
Gary B.
Manager
V, B, 4.7, 4.19
Hardware Design
Engineers
A, B, D, C, E
Software Design
Engineers
A, B, D, C, E
Process Engineers
A, B, D, C, E
Quality Documentation
Paul A.
Supervisor
B, C, 4.5, 4.16
Reliability Engineering
Sam O.
Supervisor
B, C, D, 4.2,10,11,13,14,17
Quality Control
Rosa L.
Supervisor
B, C, E
QC
Inspectors B, D
Assembly
Tony C.
Supervisor
B, C
Pick & Place
Operators
B.
General
Assemblers
B
Test
Emanual Q.
Supervisor
B, C
Maual Test
Operators
B, C
Automated Test
Operators
B, C
a
b

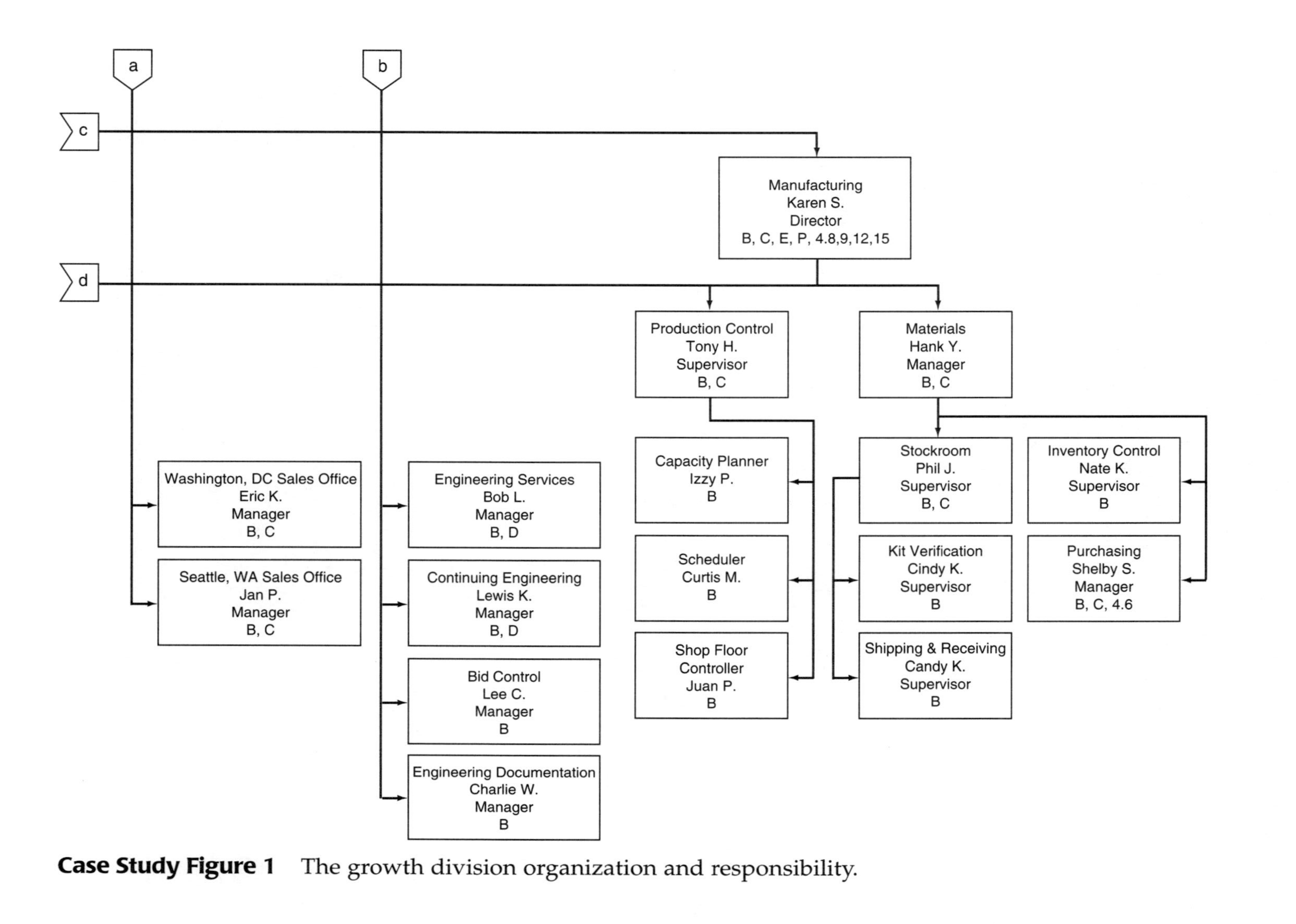

Case Study Figure 1 The growth division organization and responsibility.

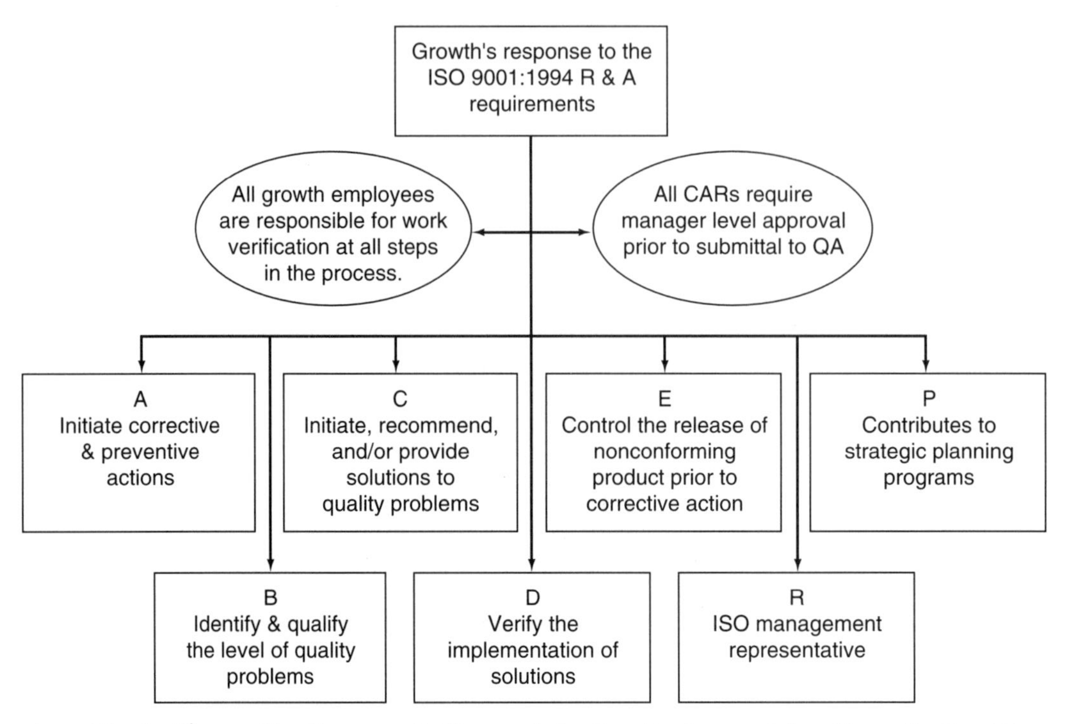

Case Study Figure 2 Responsibility and Authority Legend for Growth Quality Policy Manual Appendix C

"O.K. guys," she began, "I would like to suggest that we go ISO, and use ISO 9001 as the basis for a program to meet the 13 percent PBIT. It would give all of us a standard to measure ourselves against and would force us to learn a lot about our business—and ourselves when we are reviewed by expert strangers. Where are we on this?"

There followed an extended discussion based on what everyone knew about ISO—from both the circulated articles and their own experiences—and it seemed as if everyone had a somewhat different idea about just what the program was.

However, ISO was not completely unknown to Growth since they had already obtained a CE mark for EMC compliance to allow sales into Europe. This effort had been headed by Marlin, who was as detailed and thorough a chief engineer as you would hope to find. Janis could vouch very easily for the improved international sales that had resulted from this activity.

In addition, Mike had just completed a four-day internal ISO 9000 quality auditor training program and already had a basic internal and vendor quality audit system in place. Karen, in her normal activist mode, had supported Mike all the way in this objective and had introduced an MRB program to support the ongoing first-pass test yield studies.

Based on this set of data, in conjunction with Fran's many positive inputs on the subject, the group agreed to formally launch an ISO 9001 certification program. Mike H. volunteered to be the ISO 9000 management representative—there were no arguments, only a group sigh of relief—and his first action was to suggest the use of a quality management consultant whom he knew very well and had worked with just prior to joining Growth.

The consultant, Sam S., had over thirty years in hi-tech research, development, and operations experience and had already helped a dozen companies through the process. He was also an internationally certified quality systems lead auditor and could present up-to-date information on how the process really worked.

Fran's task now was to get clearance from Judy and the board.

Fran's Request

The financial package that went to corporate was loaded with cost-saving arguments. However, there was a new expense line entitled, "ISO 9001 Certification," and the anticipated cost reductions had to include these costs into the financial argument. They did.

What it boiled down to was that

- ISO 9000 would be the platform from which Growth would launch a concerted cost reduction program.
- ISO 9001 was ideally suited for this purpose because continuous improvement was inherent in the Standard.
- In fact, the Standard stressed customer satisfaction and was directly in line with the latest sales & marketing strategic plan.
- There was an adequate base of knowledge within the management group about ISO and its implementation.
- An extremely qualified consultant was to be hired who had been through the process a dozen times and who was also a third-party assessor who subcontracted on a part-time basis to one of the largest ISO 9000 registrars in the world.
- It was clear that several already ISO 9001–certified competitors were winning jobs from Growth because of a direct requirement for ISO certification as specified in the request for proposal's terms and conditions.
- The literature contained numerous examples of significant pay-back in cost of quality when the program was run in a serious fashion.
- Growth was to absorb all of the additional program costs—although, if successful, the program would be used as a model for the other divisions.

Board Decides

Fran received the boss's FAX to "Go ISO" at 11:35 A.M.! By 11:37, everyone in the facility had heard the news. Fran, in her heart of hearts knew, without a doubt, that the gang at corporate was completely behind her and the ISO Team.

She also knew that they would not fail, because they had carefully analyzed the financial impact of their decision and had prepared a detailed plan that assured that the necessary resources would be available when needed and at the correct level. She also know that the company had responded heroically in the past to master challenges and that they would surely respond again in a professional manner.

Of course, a bunch of significant awards and a little extra notoriety for those who achieved outstanding results would be icing on the ISO cake. As a result, Fran began to type a heads-up memo to the top managers for discussion at the next monthly review.

Requirements Analysis

The Plan

Sam S. and Mike H. spent a good many hours on the phone as they scheduled the requirements analysis assessment at Growth.

This activity consisted of two full days of on-site auditing across each department of Growth and against all twenty of the ANSI/ISO/ASQC Q9001-1994 elements. Sam's plan was quite detailed and he made sure that each member of the top management team and each product area was given enough interview time to really understand where the company was with respect to full conformance.

Even though two days was relatively short for this extensive an activity, more time would not be useful since there had been little formal development of an ISO system—just bits and pieces here and there throughout the facility. A Quality Manual did not really exist and there were no such things as process documents, although documentation at the procedural and forms level was more than adequate. The task was to determine what level of documentation made operational sense and which documents were either missing or redundant.

Sam completed his work on time and issued his report.

The Report

As indicated in the Summary Sheet on Growth (Case Study Table 2), the major gaps appeared in the areas of 4.1 Management Responsibility, 4.4 Design Control, 4.5 Document & Data Control, 4.10 Inspection & Testing, 4.14 Corrective & Preventive Action, 4.16 Control of Quality Records, 4.17 Internal Quality Audits, and 4.18 Training. As a result, the overall 4.2 quality system status was weak.

Sam then put together a schedule of events that he felt would be sufficient to bring the company to initial assessment readiness in twelve months. This Schedule for Growth's ISO Deployment is shown in Case Study Table 3.

Because of the significant effort in software development, Sam also suggested that Growth comply, where applicable, with the ISO 9000-3 software guidelines and state that the system conformed to this standard. This directive would push the group a bit, but in the long run it would help to solidify their somewhat hesitant methods of software reliability testing.

After review by the top managers, the schedule was approved and the ISO certification workshops began.

Case Study Table 2

Summary Sheet on Growth (by Percentage)

Element ISO 9001	Activity Covers Quality Manual/System	Documentation	Implementation	Demo of Effectiveness	Comments
4.1	Mgmt Responsibility	20	40	40	Weak exec review
4.2	Quality System	50	40	30	Good base in Tier III/IV
4.3	Contract Review	30	70	60	Growth team work
4.4	Design Control	20	50	50	Free running
4.5	Document & Data Control	20	50	50	Tier III/IV controlled
4.6	Purchasing	60	60	60	Good vendor control
4.7	Control of Customer-Supplied Product	20	70	70	Weak in documents
4.8	Product Identification & Traceability	60	70	70	Tier III/IV looks good
4.9	Process Control	60	70	70	Tier III/IV looks good
4.10	Inspection and Testing	40	50	50	Relies on QC to catch
4.11	Control of Inspect, Meas, & Test Equipment	60	70	70	Tier III/IV looks good
4.12	Inspection & Test Status	60	70	70	Tier III/IV looks good
4.13	Control of Nonconforming Product	60	70	70	MRB Tier III/IV looks good
4.14	Corrective & Preventive Action w/cust complaints	30	40	30	Cust cmplnts OK
4.15	Handl'g, Store, Pack'g , Preserv, & Delivery	60	70	70	Tier III/IV looks good
4.16	Control of Quality Records	40	50	20	Disorganized
4.17	Internal Quality Audits	25	25	20	Just starting up
4.18	Training	40	40	20	Seat of the pants
4.19	Servicing	60	70	60	Very responsive
4.20	Statistical Techniques	60	60	60	Good base in eng'g

CASE STUDY TABLE 3

Growth's ISO Deployment

	Months/Activities/Teams	**97 Jan**	**Feb**	**Mar**	**April**	**May**	**Jun**	**July**	**Aug**	**Sept**	**Oct**	**Nov**	**Dec**	**98 Jan**	**Feb**
1.	ISO reqts analysis w/plan														
2.	Basic ISO training all employees														
3.	Develop process documents drafts														
4.	Review process drafts & publish														
5.	Special training (flow, doc)														
6.	1st draft Quality Policy Manual														
7.	1st Quality Policy Manual edits														
8.	Final QP Manual edit & release														
9.	First edit of process docs														
10.	Final edit of process docs & release														
11.	1st draft of Tier III docs														
12.	1st edit of Tier III docs														
13.	Final edit of Tier III & release														
14.	1st draft of Tier IV docs														
15.	1st edit of Tier IV docs														
16.	Final edit of Tier IV & release														
17.	Quality auditor training														
18.	Initiate auding Program														
19.	QMS effectiveness check														
20.	Contact & select registrar														
21.	1st simulated initial assessment														
22.	Corrective action workshops														
23.	Final simulated initial assessment														
24.	Corrective action workshops														
25.	Initial assessment by registrar														
26.	Correct nonconformances														
27.	Receive certificate														

THE BUY-IN

Attendees

The first stage in the training process was for Sam and Mike to bring the top managers and other key employees up to a common level of ISO understanding. This was accomplished by means of the buy-in workshop that Sam presented in one day and which covered the basic concepts of ISO and critiqued his requirements analysis data.

In addition to Wendy G., their MIS Manager, the top management group chose to add Rebecca A., district sales manager; Rose C., head of program management; and Gary B., the customer service manager from sales & marketing.

Design engineering was represented by Bob L. Manager of engineering services; Lewis K., who headed continuing engineering; Charlie W., manager of engineering documentation; Lee C., bid control manager; Cas M., their hardware design manager; Joe P., manager of software design; and Wendell K., head of process engineering.

Quality assurance was represented by Paul A., who supervised quality documentation; Rosa L., Supervisor of quality control; and Sam O., head of reliability engineering.

The top management group also chose Tony C., the assembly supervisor; Emanual Q., test supervisor; Tony H., head of production control; Hank Y., the materials manager; and Shelby S., purchasing manager, from the manufacturing department.

The three administrative assistants, Anne S., Kathy L., and Brenda W.—to the VP & general manager, to sales & marketing, and to engineering, respectively—completed the set of trainees.

In this manner, each major function of the company was represented and the training would lead to a common ISO language.

Activities

After a review of the three pillars of ISO 9000—documentation, implementation, and demonstration of effectiveness—and a discussion of the ISO 9000 schema, Sam presented the basic concepts behind the design of an effective ANSI/ISO/ASQC Q9001-1994 quality management system.

During this process he reviewed all twenty elements of the Standard and indicated where ISO 9000-3 could be used to enhance Growth's total program. In addition, Sam reviewed the strengths and weaknesses uncovered in the requirements analysis and presented the schedule for Growth's ISO certification.

The group was obviously excited about the aggressive schedule, although several members voiced their anxiety about an initial assessment at the close of the fiscal year. December was usually a go-like-the-devil month to make the numbers. Obviously, they had all better learn to plan their booking and shipping rates with more care. Perhaps ISO would help in this area?

Sam had begun the session at 8:00 A.M., and by 3:00 P.M. the group, which had been active and truly concerned with the program, was now obviously restless and a little weary of so much theory. The use of the cafeteria and its plastic seats, needed to contain this large group, only added to their discomfort, especially for those with little meat on their bottoms. Most of the attendees had never stayed in a class that lasted more than 50 minutes, so this was obviously a stretch. At this point, Sam decided to skip a few overheads and launch the stewardship exercise.

Fran and her direct reports had been advised of this exercise weeks before and had formed opinions on who would be chosen for the various posts, but it was best to do it as a group and get the group's formal acceptance and commitment for this critical task. Sam began with 4.1 Management Responsibility and worked all the way down to 4.20 Statistical Techniques. It took over an hour, and at several points the group exploded in laughter as a favorite foil was socked with an assignment. The exercise did its usual job in allaying ISO fears and alloying the group into a common goal—get certified and get this bloody thing over with.

When Sam carefully explained that after the initial assessment the spirits of ISO return every six months, or at most every year, for surveillance, there was a good-natured collective groan. Indeed, this was the first step in a continuing journey to enhanced profitability and productivity from which all the attendees would eventually benefit.

The stewardship decisions are shown in (Case Study Table 4) "Go ISO" had suddenly become a reality, and each attendee walked out of the cafeteria ready for action but not quite sure about what that meant. The first documentation workshop would tell them.

CASE STUDY TABLE 4 **Growth's Stewardship Assignments**				
Ref#: Initial **Date(s): 15 Jan 97**	**Supplier Name/Location:** **The Growth Division**		**Type: ANSI/ISO/ASQC** **Q9001–1994**	
Element ISO 9001	**Activity Covers Assignment of Stewards & Continuous Improvement Managers**	**Stewards by Name**	**C/I Program Managers by Name**	**Team Members**
4.1	Management responsibility steward Management representative	Fran L.	 Mike H.	Greg C., Janis K., Marlin P., Karen S., Mike H.
4.2	Quality system steward	Mike H.		Charlie W.
4.3	Contract review steward	Janis K.		Fran L., Rebecca A., Brenda W., Gary B.
4.4	Design control steward	Marlin P.		Cas M., Joe P., Wendell K., Lewis K., Tony H.
4.5	Document & data control steward	Paul A. & Charlie W.		Anne S., Wendy G., Kathy L.
4.6	Purchasing steward	Shelby S.		Lee C., Bob L.
4.7	Control of customer-supplied product steward	Gary B.		Lewis K.
4.8	Product ID & traceability steward	Karen S.		Tony C., Emanual Q., Tony H.
4.9	Process control steward	Karen S.		Tony C., Emanual Q., Tony H.
4.10	Inspection and testing steward	Mike H.		Sam O., Tony C., Tony H.

Case Study Table 4
Continued

Element ISO 9001	Activity Covers Assignment of Stewards & Continuous Improvement Managers	Stewards by Name	C/I Program Managers by Name	Team Members
4.11	Control of I M & TE steward	Mike H.		Cas M., Joe P., Sam O.Sam
4.12	Inspection & test status steward	Karen S.		Tony C., Emanual Q.
4.13	Control of nonconforming product steward	Karen S.		Sam O., Rosa L., Karen S., Hank Y.
4.14	Corrective & preventive action & customer complaints steward Preventive action C/I manager Customer complaints C/I manager	Mike H.	Gary B.	Fran L., Geg C., Marlin P.
4.15	Handl'g, storage, packaging, pres & del'y steward	Karen S.		Rosa L., Tony C., Emanaul Q., Tony H., Hank Y.
4.16	Control of quality records steward	Paul A.		Anne S., Wendy G., Kathy L., Brenda W., Charlie W.
4.17	Internal quality audits steward Internal quality audits C/I manager	Mike H.		Anne S., Wendell K.
4.18	Training steward Training C/I manager	Greg C.		Fran L., Janis K., Marlin P.
4.19	Servicing steward	Gary B.		Fran L., Janis K.

CASE STUDY TABLE 4 Continued				
Element ISO 9001	**Activity Covers Assignment of Stewards & Continuous Improvement Managers**	**Stewards by Name**	**C/I Program Managers by Name**	**Team Members**
4.20	Statistical techniques steward	Marlin P.		Lewis K., Sam O., Karen S., Shelby S.

Stewards Manage "channels" of information, that is, policy, process, procedure, and forms, to ensure that the channel is fully documented, implemented and demonstrating effectiveness in meeting the organization's quality objectives.

C/I Managers Formulate, implement, and effectively manage their specific programs, for example, the preventive action program, the audit program, the training program.

Approved: Fran L. **Dated: 15 January 97**

GROWTH'S PROCESSES

Techniques

Sam preferred to use a narrative style for the Quality Policy Manual; flow charts for the Process Manual; and tables for the Tier III procedural documents.

By the use of narration, Growth's personality would become clear to the unusually large set of diverse Quality Policy Manual readers; the flow chart technique for the Process Manual permitted a relatively quick overview of complex interdepartmental functions; and procedural tables avoided the sometimes inconvenience of flow-chart loops that could be misread by weary eyes. It was strictly a matter of preference, and any style could be made to work with a little patience and a solid training program.

To prepare for this exercise, Sam held training sessions with the three administrative assistants, the engineering documentation manager, the quality documentation supervisor, purchasing manager, and the MIS manager to ensure that there was sufficient expertise available for both flow charting and procedural table creation. He also encouraged the group to take outside courses on these subjects, which they eventually did.

Processes

The first documentation workshop stressed the creation of process flow charts, which totally described every aspect of the entire Growth operation—from marketing through after-sales service. Sam worked diligently to expedite the process, and by the end of the allotted three-day workshop all the stewards were off and running.

The stewards were required to complete their first drafts within thirty days so that a first-shot Hub document master list could be created and that there would be adequate research data available to create the Quality Policy Manual. The work groups were instructed to either write out their processes directly into flow charts or to use a narrative that could then be transformed by the selectively trained group into flows.

After a few weeks, Sam held the follow-up two-day critique with the stewards and was pleased that they had responded so well to the challenge. Although the first dump actually took 47 days to complete, due to normal business pressures, the first Hub document list was published (Case Study Table 5). An example of Growth's process flow charting protocol is shown in Appendix F.

Case Study Table 5

Growth's Hub Documents

ISO 9001 Element	Manual Section	Hub Documents
4.1	One	▪ Stewardship summary ▪ Operating budget
4.2	Two	▪ Tier I—Quality Policy Manual ▪ Employee handbook ▪ Master forms catalog ▪ Standards & codes procedure ▪ Technical publications manual
4.3	Three	▪ Contract review process ▪ Growth's price list
4.4	Four	▪ Hardware & software development guidelines
4.5	Five	▪ Document & data control guidelines
4.6	Six	▪ Purchasing Manual
4.7	Seven	▪ RMA process
4.8	Eight	▪ Hardware & software manufacturing processes
4.9	Nine	▪ Hardware & software manufacturing processes
4.10	Ten	▪ Inspection & testing processes
4.11	Eleven	▪ Metrology Manual
4.12	Twelve	▪ Hardware & software manufacturing processes
4.13	Thirteen	▪ Control of nonconforming product process
4.14	Fourteen	▪ Corrective & preventive action process ▪ w/customer complaints
4.15	Fifteen	▪ Hardware & software manufacturing processes
4.16	Sixteen	▪ Quality records procedure & master list
4.17	Seventeen	▪ Quality audit processes
4.18	Eighteen	▪ Training program ▪ Job descriptions
4.19	Nineteen	▪ Servicing process ▪ Terms & Conditions Manual
4.20	Twenty	▪ Data & statistical analysis processes

Growth's Quality Policy Manual Formats

Format

Growth's Quality Policy Manual (Manual) was designed to demonstrate conformance to all twenty of the Standard's "shalls," as well as respond where applicable to the ISO 9000-3 software guidelines.

The format chosen was "stand-alone," and it covered all twenty elements in sequential form and used the Standard's titles to define each section.

At Sam's urging, the entire policy, process, and procedural documentation started out as hard copy and would eventually be transformed into on-line documents once confidence had been reached in the efficacy of the system. Wendy had already begun to calculate the cost of the required intranet.

The narrative style was embellished somewhat with a documentation hierarchy pyramid in "Section 2, Quality System," and a simplified graphic for the design and manufacturing protocols in "Section 4, Design Control," and "Section 9, Process Control," respectively.

Cover Pages

As a stand-alone document, the Manual's cover page clearly defined the purpose of the Manual and the VP & General Manager's approval. There was a statement to the effect that, "*This is a controlled document. Copies of such documents may not be made or distributed to any person or persons, companies, or organizations, without strict adherence to both* Growth's *document control procedures and disposition of proprietary information.* Growth *reserves the right to change, modify, and/or withdraw this document, without notice, other than under those conditions specified by the ISO 9000 registrar.*"

A control number in "red" was placed in the right-hand top corner of the cover page to signify control. The VP & general manager, as the document's owner, approved the release of uncontrolled Manuals to any one that either the director of sales & marketing or the purchasing manager also approved.

The numerical descriptor for the Manual was placed in the upper left-hand corner, QA-1-02-001-01 (responsible department—Tier I document—in response to ANSI/ISO/ASQC Q9001-1994 element 4.2—the first document in the series—revision level 1). Just below the numerical descriptor was Paul A.'s name as the document issuer.

The next set of pages covered document control activities, the origination date with a requirement for annual review, and a table of contents. The table of contents stated clearly that, *"The above denoted sections and titles refer directly to the pertinent ANSI/ISO/ASQC Q9001-1994 elements in the Standard."*

Body Style

The Manual started out with a description of Growth's business and its vision, mission, and quality policy statement to capture the concept of an integrated business and quality strategic declaration. Each section of the Manual was written by a subject matter expert—who were generally the stewards, with review and editorial support from their teams.

There were a few mandatory rules, for example, each section must state responsibility for the activities addressed and refer to the appropriate Hub document, and proprietary information was not to be discussed. Otherwise, the style was free and easy and open and designed to suit the writer's personality and knowledge of the area.

Sam spent a series of two inclusive three-day reviews on the Manual with the stewards and their teams, and gave each steward a portion of the time commensurate with the section's complexity. The rule of thumb was that each team needed to spend four times the time he spent with them during the drafting phase. The teams responded extremely well, and the Manual's first draft was completed within the allotted 90-day window.

Without the intense work performed on the Process Manual, this exercise would have taken twice as long. As a result, within four months Growth had Tier I and Tier II at the first-draft level.

We will now describe the specific manner in which the stewards and their teams created the Manual.

MANAGEMENT RESPONSIBILITY POLICY

Method

To understand how Growth's Stewards created the Manual, we need to respond to each requirement of the Standard in its turn. The exact order of discussion could be an issue, but the technical group that defined the Standard did so in such an efficient manner that there is really no sense in inventing a different sequence within each element. We will have an intelligible and useful document if we simply respond to each "shall" as it appears.

Accordingly, we will take each "shall" in its turn and describe Growth's response to the "shall" in terms of quality policy statements.

The gist of the Standard's actual text is shaded and in a smaller font, and appears first. Growth's analysis by the steward and team and subsequent response appears second and is shown in normal font. The Manual's text is shown indented in quotes, that is "Manual."

Alert

Should the reader wish to, the extraction of all "quoted text" would form the complete text of a Quality Policy Manual.

We begin with element 4.1 of the Standard (Management Responsibility). Growth, in their response via quality policy statements, has labeled this work, "Section 1 Management Responsibility." They then kept to this pattern, e.g., "Section 2 Quality System," "Section 3 Contract Review."

Section 1
Management Responsibility Policy

Fran carefully examined the first clause of element 4.1, i.e., "4.1.1 Quality Policy," and mouthed the words to make them sink deeply into her mind.

4.1.1 QUALITY POLICY

Growth's top management will need to define and document its overall quality policy, objectives, and commitment to quality.

Fran's Response

Fran had decided to write the quality policy statement for "4.1" and then preview her text with the top managers. This section was far too important to leave for anyone else because it clearly defined Growth's vision and path to success. She was impressed with Sam's integrated approach to quality and business as inseparable objectives. For the first time she really sensed what ISO was all about.

She began with a brief history of Growth's achievements and then described the scope of the certification:

"Growth as a Business Profile

The Growth Division was established in 1993 as an operating group of the Stable Corporation, a business enterprise founded in 1985, with corporate offices in Dallas, Texas. Growth has been charged by Stable Corporation's board of directors to develop and market hardware and software accelerator boards for data acquisition purposes across a very wide range of commercial, medical, and military applications.

Growth's first series of microlayered board, box, and subsystem components was very well received by OEMs, and this allowed the company to approximately double its sales each year for the past four years. Growth's movement toward ANSI/ISO/ASQC Q9001-1994 certification is indicative of the division's desire to continue this impressive growth in concert with increased quality and with customer satisfaction foremost in mind.

Scope

The Growth Division designs, manufactures, markets, sells, and services modular hardware and software products as a means to display and process industrial and medical images on personal computers and workstations."

At this point, Fran felt that she had addressed the initial set of "shalls" except for the quality objectives. For this section she wanted to stress the transparency of Growth's business and quality objectives. She began with the division's vision.

"Vision

Growth's vision is to be the leading supplier of microlayered board, box, and subsystem technology for imaging systems.

Mission

To achieve this goal, Growth will

- Work on a partnership basis with customers to satisfy their technological needs.

- Design products that meet the customers' explicit and implicit requirements.
- Manufacture products that are delivered against the customers' on-time requirements.
- Respond quickly and thoroughly to customer complaints and service requests.
- Maintain an effective ANSI/ISO/ASQC Q9001-1994 quality management system that also complies with the applicable clauses of the ISO 9000-3:1991 software guidelines.
- Meet its financial goals in agreement with corporate requirements.
- Provide a responsive and rewarding work environment for its employees.

Objectives

To satisfy our mission requirements, Growth will

- Periodically survey our customers to establish both satisfaction and dissatisfaction levels.
- Follow a strict regimen of hardware and software design reviews.
- Track first-pass test yields and returned product rates.
- Carry out an intense program of vendor/subcontractor evaluation.
- Perform an extensive activity in corrective and preventive action and customer complaint response.
- Closely monitor our financial goals."

At this point, Fran was satisfied that she had fully addressed the first set and now looked closely at the next set of "shalls."

> The quality policy was to be relevant to Growth's organizational goals and the expectations and needs of its customers.

By examination, Fran was positive that the text before her had already addressed this clause and decided to move forward to the next set.

> Growth is to ensure that the quality policy is understood, implemented, and maintained throughout the organization.

Fran mused over this requirement for a bit and then realized that there were a number of ways that top management communicated their policies to the employees. As a response, she decided to simply list them.

"Communication

Growth uses several means to propagate its quality policies to all employees. Aside from the monthly top management meetings, such methods include

- The assignment of stewards who establish quality teams to document, implement, and demonstrate the effectiveness of the quality management system (by implementation of "Growth's Stewardship Summary" maintained by the ISO 9000 Management Representative).
- Weekly business review meetings by each top manager with their staff.
- The division's monthly newsletter.
- Quarterly business presentations by the VP & general manager to all employees."

As far as Fran was concerned, this did the job. It was on to the next set, which looked like quite a challenge. It was.

4.1.2.1 Responsibility & Authority

This section required a detailed discussion of how responsibility and authority is managed within Growth in terms of those who manage, perform, and verify work.

Fran was a little stumped at how to handle this one, but she remembered Sam's admonition to keep it complete but simple. As a result, she dug out the training workbook and found that Sam had suggested the use of an organization chart and matrix of responsibility and authority to handle this clause.

A sample of the way Fran resolved her dilemma were shown in the Case Study Figures 1 and 2, entitled "The Growth Division Organization and Responsibility," and "Responsibility and Authority Legend for Growth," respectively.

The legend responds directly to the conditions 4.1.2.1 a) through e) in the Standard.

Fran typed into the computer

"Responsibility and Authority

The following organizational charts indicate the functional relationships of all personnel in Growth, along with the indicators that summarize which employees have the responsibility and authority to manage, perform, and verify work affecting quality. The legend for the indicators follows directly after the organizational charts.

In regard to verification, all employees of Growth are required to constantly monitor their work to ensure that all quality requirements have been satisfied. In addition, the series of numbers from 4.1 through 4.20 relate directly to the Standard and indicate ISO 9000 stewardship whereby each steward is responsible for the effective documentation, implementation, and demonstration of effectiveness for his or her element.

The specific responsibilities for Growth's top management team is discussed after the charts.

Direct Report Management Responsibilities

VP and General Manager

Responsible for developing and overseeing the execution of Growth's annual operating budget and managing the successful implementation of that budget. As the top divisional executive, the VP & general manager is required to articulate Growth's long-term strategies and to serve as the primary interface between the division and the corporate offices.

Director, Sales and Marketing

Required to achieve the annual domestic and international sales quotas and direct the efforts of the sales personnel, program managers, customer service, and the Washington, DC, and Seattle, WA, sales offices. The position also requires the creation and implementation of the marketing plan, which includes media contact, direct mail, trade show activities, and OEM contacts.

Director, Design Engineering

Serves as the chief engineer for Growth and is responsible for the development of the division's product line. This activity includes synthesizing customer requests for product development and the general oversight of all aspects of hardware and software design functions. Process and continuing engineering as well as engineering services, bid control, and engineering documentation are directly controlled through this position.

Controller

Creates and generates all financial information for internal and external users, and is responsible for financial planning and management of the accounting functions. The controller is the primary interface with the CFO at Outstanding.

Director, Quality Assurance

Serves as the ISO 9000 management representative for Growth and manages all phases of quality assurance, quality control, and reliability engineering. The position also requires the oversight of the policy, process, procedural, and format document control for Growth.

Director, Manufacturing

Responsible for the management of the production, materials control, and all purchasing for Growth. This position includes the functions of inventory management, the evaluation of vendors, and the support of prototype product builds.

MIS Manager

Responsible for the design and implementation of all phases of the information technology system at Growth and is the primary contact with CITO at Outstanding. The MIS Manager controls the day-to-day effectiveness and back-up systems for the division."

Although the section appeared to be somewhat lengthy, Fran remembered Sam's admonition to tell the story that is necessary. The length will take care of itself. The effective response to 4.1 requires a good amount of discussion; it cannot be avoided.

Fran quit for the night and started out early the next morning to type out the next set of "shalls."

4.1.2.2 Resources

Growth is to identify resource requirements and provide adequate resources to its employees to meet the demands of an effective QMS.

This one seemed easy for a change, and she literally dashed off her response. Apparently, all the preparation required to get to this clause had loaded her fingers with clarity.

"Resources

Growth is required to compose and implement an annual operating budget at both the top management and second-tier management levels. This budget is maintained by the controller and is used to generate a personnel hiring plan, the capital spending plan, and detailed operating budgets for each department at Growth.

During the fiscal year, top management reviews the budget on a monthly basis and compares the division's performance to the plan. The plan is then adjusted as required.

Key elements of this plan include an extensive training program for all employees including all aspects of management, hardware and software development, and internal quality auditing."

With this section completed, Fran could breathe a little easier because she was now past the halfway point and headed for home. She grabbed a diet drink and attacked the next section.

4.1.2.3 Management Representative

Growth is to select and document the duties of its ISO 9000 management representative.

"Management Representative

The VP & general manager of Growth has appointed the director of quality assurance as the ISO 9000 management representative. Notice of this appointment was effective on the first day of February 1997, and was distributed to all employees of Growth. Such duties are in addition to the usual activities of the quality assurance director.

In this position, the representative has the authority to establish, implement, and maintain an effective ANSI/ISO/ASQC Q9001-1994 Quality Management System within Growth, subject only to the review of the VP & general manager and the division's directors.

The duties of the ISO 9000 management representative include the establishment and management of our plan to meet the established certification timelines and to provide any necessary direction to the ISO stewards.

It is also required that the representative report at the management review meetings on the status of the ISO Readiness Program in the areas of at least

- Progress against our goals and objectives
- Internal quality audits
- Corrective & preventive actions & customer complaints
- Training
- The state of documentation & implementation of the quality system

so that the appropriate actions can be taken to improve the system.

The representative will also Interview and choose an accredited registrar, during the final phase of our readiness program.

In addition, the representative will ensure the efficacy of the quality management system by directly taking part in the internal quality audit and corrective and preventive action programs within Growth."

Fran also released a memo to this effect that directly supported the Manual's language. The memo read

To: All Growth Employees
From: Fran L.
Dated: 1 February 1997

To ensure the success of our new ISO 9000 certification effort, it is my pleasure to appoint the director of quality assurance, Mike H., as the ISO 9000 management representative for Growth, effective immediately, and irrespective of Mike's other duties.

In this position, Mike will have the full support of me and my staff and the authority to establish, implement, and maintain an effective ANSI/ISO/ASQC Q9001-1994 quality management system.

Mike's duties as ISO 9000 management representative, require him to

1. Establish and manage the plan to meet our goal of certification by the end of 1998.
2. Provide top management overall direction to the ISO stewards and program managers.
3. Activate a program designed to ensure the propagation and implementation of our ISO quality policies to all levels of Growth.
4. Report to the management review team on the status of the ISO Readiness Program in the areas of at least
   - Progress against our goals and objectives
   - Internal quality audits
   - Corrective & preventive actions & customer complaints
   - Training
   - The state of documentation & implementation of the quality system
5. Based on such reviews, determine the suitability and effectiveness of our quality management system and initiate any changes necessary to continuously improve the system.
6. Provide the conduit whereby key information from other operational meetings, which take place outside of the management review, flow into the main review.
7. Interview and choose an accredited registrar, during the final phase of our readiness program, to take us through the initial assessment and be our partner as we optimize the system via their surveillance visits.

Please support Mike in this new challenge, so that we can perform a highly successful ISO certification effort.

Fran felt satisfied that she had created a fully compliant response to the Standard and was now prepared to address the next clause.

4.1.3 Management Review

Growth's top management is to hold extensive management reviews to ensure the viability and efficacy of the QMS.

Again, the previous exercises had set the stage for Fran's quick response. In addition, Sam had advised her to address the supplier/customer joint software review requirement from ISO 9000-3 and she immediately typed out

"Management Reviews

Quarterly

The ISO 9000 management review is held quarterly and is chosen from one of the monthly top management meetings attended by the VP & general manager and the directors.

At this meeting, the total business performance is reviewed and the suitability and effectiveness of the ANSI/ISO/ASQC Q9001-1994 quality management system is determined and the necessary actions taken to improve its performance.

The agenda for the review is set by the VP & general manager and includes, but is not limited to, the total corrective and preventive action and customer complaints program, including an analysis of nonconformities; the total auditing program; a review of our quality objectives as compared to plan; and the training program. The presentation of such data is by the appropriate attendees.

The minutes for this meeting are written and maintained by the administrative assistant to the VP & general manager.

Supplementary Reviews

Monthly top management reviews and weekly departmental level reviews are also held as a way to analyze the effectiveness of the system on a much finer grid than over three months. Minutes of such reviews are discretionary, but the key information collected at such meetings is funneled into the quarterly review, as appropriate.

Joint Software Reviews

Growth's software development protocol requires a close interaction with its customers in the form of jointly reviewed conformance to customer specifications. Conformance is based upon software acceptance testing at both Growth and the customer's facility. The director of design engineering schedules and manages this activity."

This response brought Fran to the end of the element and she dashed off a memo to the directors requesting them to review and comment on her creation. Aside from the need for some cleanup, she felt it was a good piece of work. So did the directors.

QUALITY SYSTEM POLICY

Response

Mike took about the same position as Fran when it came time for him to write his section on the quality system. Because of the universality of the topic, he felt it would be best to write a first draft and then review it with the team.

The process flow chart exercise and subsequent Hub document master list had provided him with an unusually strong basis to summarize Growth's entire QMS. In addition, he made sure that the specific "standards & codes" and divisional interface conditions required by the registrar were also covered.

He started to type at about 7:30 on a Tuesday morning and completed his draft in the early afternoon. His composition is shown below in the same manner as we have shown Fran's in the previous section; the gist of the actual Standard's text is given first and Mike's response in bold follows directly after. Sam's suggestions on ISO 9000-3 compliance was also included in the first draft.

Aside from minor corrections, the team was extremely satisfied with Mike's work.

Section 2
Quality System Policy

4.2.1 GENERAL

Growth is to establish, document, and maintain a quality system.

"Quality System

The Growth Divison of the Stable Corporation has established a quality management system (System) is to ensure that all of Growth's products are manufactured within an environment that conforms directly to the Standard.

Life Cycle

The System is designed to impact the entire life cycle of Growth's hardware and software products. As a result, the product plans are based upon the effective

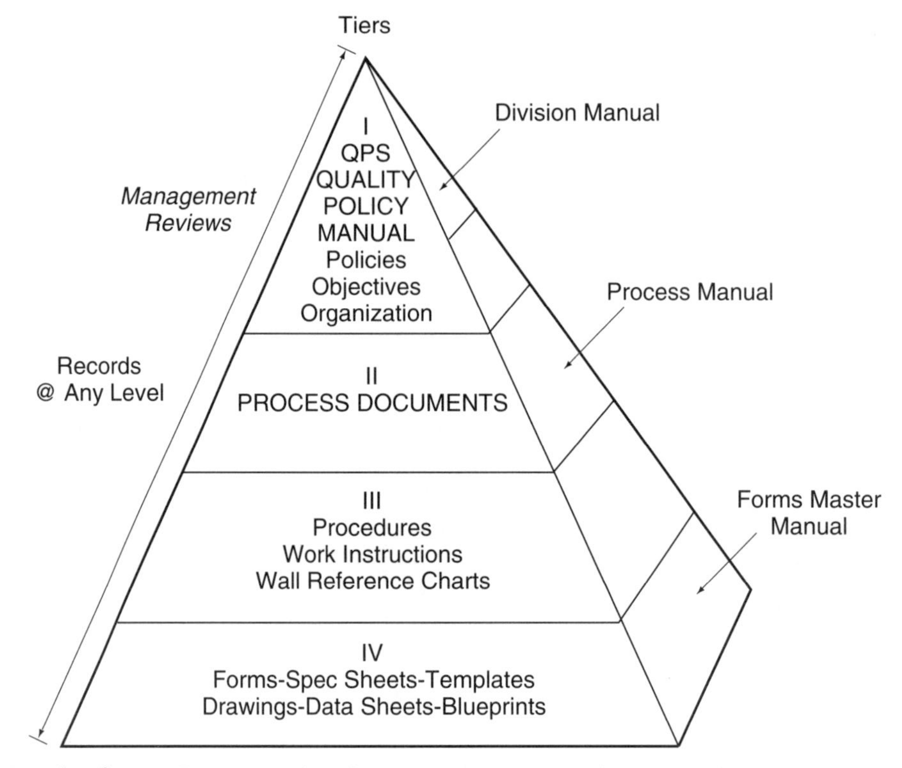

Case Study Figure 3 Growth's four-tier operational pyramid.

inclusion of all aspects of the product's life, that is, from market share to after-sales service.

The System documentation is primarily hard copy in the form of manuals for the upper-tier policy/process documents; and in the form of individual procedures/work instructions and formats/templates for the lower-tier documents. An MRP system is used by manufacturing, and engineering maintains a separate on-line documentation system.

The four-tier structure for the System is illustrated in Case Study Figure 3."

> Growth is to prepare a quality manual, and the manual is to reference quality-system procedures and outline the structure of the documentation used in the system. Guidance on quality manuals is given in ISO 10013:1993.

"Quality Manual

As illustrated in Case Study Figure 3, Growth's Quality Policy Manual (Manual) is the highest-level document in the System and defines Growth's quality policy statements for all twenty elements of the Standard and those portions of ISO 9000-3 that are applicable.

The VP & general manager is responsible for the review and approval of the Manual. As with all Growth documents, the Manual is reviewed and updated either upon revision or during the quality audit process.

Linkage

Linkage from document to document within the System is by means of referenced documents. In this manner, the reader is directed from the Manual to the process documents and thence to procedural and format documents, as appropriate.

For ease of use, the documents are coded by placing the numerical descriptor in the upper left-hand corner, that is, for the Manual, the code is QA-1-02-001-01 which indicates the

- Responsible department (QA)
- Tier level (Tier I)
- ANSI/ISO/ASQC Q9001-1994 element it responds to (4.2)
- Number of the document within that series (001)
- Revision level (1)

To expedite navigation, it is always advisable to begin with the Manual and then go directly to the Appendix, entitled "Growth's Master List of Hub Documents." The Hub document can be likened to an airport hub. Once you reach the Hub document, it leads the reader to the next levels of system information."

4.2.2 Quality System Procedures

Growth is required to prepare documented procedures and effectively implement the quality system and its documented procedures.

"Procedures

It is important to note that Growth's "procedures" are in the form of either very high-level process documents or lower-level procedures or work instructions.

The high-level process documents are equivalent to standard operating procedures and are in the form of flow charts with a supplemental text document attached. The two documents form a single process document.

Implementation

The effective implementation of the system is ensured through a comprehensive management review defined in Section 1 of this Manual. Most importantly, great care is taken to track all preventive actions achieved by Growth employees and to reward such activities commensurate with their contribution to the overall division's productivity."

Growth is to consider the range and detail of the documents used in the sense that they be dependent upon the complexity, methods, skills, and training of its processes and personnel, as they apply.

"Skill Levels

All of Growth's documents are created to serve highly skilled and extensively trained employees. Moreover, Growth's employees are required to work for lengths of time without close supervision and to carry out multitasking work.

As a result, the level of detail in the documents varies from engineering guidelines to detailed test protocols based on the specific operational and administrational tasks. Most importantly, the documentation is designed to support minimal supervision by being readily available yet unobtrusive."

4.2.3 Quality Planning

In this section, Growth is to define and document how the requirements for quality will be planned for and met in all processes presented for certification.

"Quality Planning

Growth's VP & general manager is responsible for the annual development and publication of the operating budget, as discussed in Section 1 of this Manual.

In lieu of quality plans, Growth uses a hierarchical documented system of policies, processes, procedures, and forms to control the quality planning activities. For example, software planning includes—but is not limited to—design documents, functional specifications, and quality documents to verify that the functional specification has been met.

Procurement

The acquisition of major capital equipment, new processes, and specially skilled employees is the responsibility of department managers and requires approval by the VP & general manager.

Compatibility

Growth employs design and documentation reviews to ensure that new products are manufacturable. A combination of program management/project management teams during design and the use of a continuing engineer during the production start-up phases ensures the effective transfer of new products into manufacturing. Process engineers are then used to maintain the production lines.

Installation

Installation guides are provided to customers so that they can effectively install Growth products into their systems. Developer's guides provide information for developers writing software applications and software libraries such microsoftware plus have resident software library references.

Updating

Design engineering in conjunction with quality assurance is responsible for updating, as necessary, quality control procedures, inspection and testing techniques, and the development of new test instrumentation.

Design engineering has the primary responsibility to design and implement any measurement equipment that exceeds state-of-the-art specifications. Such equipment must be available prior to production release and be clearly identified in the project plans.

Verification and Validation

Design Engineering is primarily responsible for the functions of verification and validation at scheduled stages in the product design plan. Upon release of the product to production, quality assurance imposes verification and reliability activities on the product at all stages of manufacturing.

Standards, Codes, and Workmanship Standards

The "Standards & Codes Procedure" lists all of the regulatory and statutory requirements imposed on Growth's products, for example, ISO 9000 Standards and applicable guidelines, IPC workmanship standards, and the CE mark. The procedure includes a master list that defines the standards and codes used, the responsible employee, how the codes are kept current; and where they are located. The internal design standards are also included in this procedure.

Workmanship standards in the form of a series of IPC documents and reference manuals are provided to manufacturing. All manufacturing personnel are required to attend and satisfactorily complete a certification program based on IPC standards.

Records

A master list of quality records is maintained by the director, quality assurance to ensure that each department adheres to the Standard's requirements. The master list clearly denotes where the originals are stored and the employee who is responsible for their maintenance. The retention times noted in this list are approved by the controller."

Corporate and Interdivisional Interfaces

The Growth Division interfaces with other Stable corporate facilities. In those cases where Growth provides services to other facilities, the transactions are performed exactly as if Growth were selling its services to a customer.

In those cases where Growth receives services from another facility, the transactions are performed as if Growth obtained material from a vendor or subcontractor. In addition, the functions of financial analysis and information technology are shared directly with corporate. In such cases, the transactions are covered in procedures controlled by Growth."

CONTRACT REVIEW POLICY

Response

Janis convened the entire 4.3 contract review team and analyzed the detailed process flow chart that had been previously prepared for this activity and which now was listed as a Hub document.

She preferred to write the section on contract review as a team because she felt that the future of Growth depended heavily on sales & marketing's ability to clearly understand Marlin's technology. It was the sales presentation of the new boards that was of concern, and she believed that if Marlin were a direct contributor to the marketing and sales protocols the procedure would ensure the technological clarity.

As a result of this cross-functional team approach, several significant revisions were made in those areas where engineering was required in the review and acceptance process. The actual composition still required an examination of each "shall," but instead of a direct input into a computer an easel was used, and the results were edited in real time and then clerically entered into the computer after the meeting. The results of the team effort are shown in "Section 3."

The draft was also reviewed by Sam, and he indicated that the pertinent areas of ISO 9000-3 were indeed covered.

4.3.1 GENERAL

Growth is to establish and maintain documented procedures.

"Section 3
Contract Review Policy Procedure

The procedure for contract review is contained in the document entitled "Contract Review Processes." The director of sales & marketing is solely responsible for the content and accuracy of this document."

4.3.2 REVIEW

Growth is to review and resolve any issues on all bids and contracts before either submission or program initiation.

"Forecasts

The director of sales & marketing prepares a rolling monthly forecast for review with the top managers to ensure both hardware and software product availability and to meet contract or accepted order specifications.

Standard Products

The VP & general manager has final review authority and the director of sales & marketing publishes Growth's standard price list.
Standard off-the-shelf products are quoted by the sales staff by discounting the published price lists. Any nonstandard discount requires the approval of the director of sales & marketing.

Custom Products

Custom quotes and customer contracts, which require special product or pricing changes, are reviewed and approved by the top management team to ensure that Growth has the proper financial, marketing, engineering, quality assurance, and manufacturing expertise to take on the project.

Customer Specifications

Growth's sales staff, either at corporate headquarters or in the sales field offices, is required to review all written and verbal quotes with the customer and/or prospect base to ensure that customer specifications are clearly addressed.

New Products

New opportunities for products—either hardware or software—solicited either by internal referendum or from the external market, are reviewed by the director of sales & marketing and presented to the top management team to make the final decision on acceptability. The director of design engineering cannot accept a new project without this review and approval.

Verbal Orders

Before a verbal purchase order can be accepted, the order must be documented by the sales staff and, as with any written purchase order, the staff must ensure that the product specifications and pricing are correct.

Conflicts

Growth's sales staff, both corporate and in field sales, have the responsibility and authority to resolve any issues in contracts and to resolve order discrepancies, and to raise such issues to whatever level of authority is required."

4.3.3 Amendment to Contract

Growth is to effectively manage contract changes.

"Amendments

Sales & marketing personnel have the sole authority to amend orders regarding changes to written contracts. They must notify all of the affected departments in writing and manage any subsequent activity, particularly when a new product effort is involved."

4.3.4 Records

Growth is to keep records of contract reviews.

"Records

Quotes are dated and stored by the sales staff. If a sales manager for a specific region is unavailable, another member of the staff has the authority to resolve any issue that might arise. All pertinent verbal and written correspondence with Growth's customers is dated and stored locally."

Design Control Policy

Response

Marlin followed Janis's approach and also convened his 4.4 design control steward's team to compose this section as a group activity.

The group carefully considered the various ISO 9000-3 requirements as specified by Sam, and did a cross-check of their work with the process document entitled "Hardware and Software Development Guidelines."

The group took nearly an entire day to complete the first draft, but in the end they felt it had captured the spirit of Growth's intense and dynamic design protocols. Their work is shown in "Section 4."

4.4.1 General

Growth is to establish and maintain documented procedures.

"Section 4 Design Control Policy

Design Flow

Growth's overall product design protocols cover six stages and this concept is shown graphically in Case Study Figure 4.

Procedures

Growth maintains documented and control procedures throughout the life of the product. A detailed description of each stage in the process is located in the Hub document entitled, "Hardware and Software Development Guidelines."

4.4.2 Design and Development Planning

Growth is to prepare and maintain plans for each design and development activity.

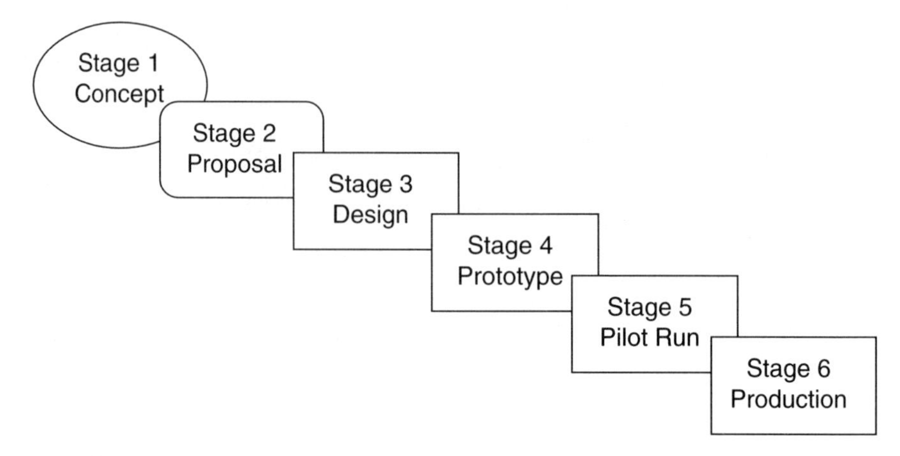

Case Study Figure 4 Product design protocols.

"Product Management

Products are designed under a program manager and project engineer protocol. The program managers are assigned by the director of sales & marketing and the project engineers by the director of design engineering. The project engineers are assigned to programs based on their expertise in hardware or software, and each project engineer is required to form a cross-functional program team.

Program Plans

The team creates the program plan and the program manager is responsible for all administrative activities, which include maintenance of the program plan and the program files. The director of design engineering, in the role of chief engineer approves the final plan after an initial design review.

Resources

The director of design engineering responds to requests by the program manager for the necessary engineers, equipment, facilities, and support personnel.

Updating

The program manager updates each design plan periodically based on design review, management reviews, or status meetings."

4.4.3 Organizational and Technical Interfaces

Organizational and technical interfaces between different groups that input into the design process are to be defined in terms of communication flow.

"Communication

The program manager schedules weekly team reviews to monitor the use of program resources and to maintain a companywide perspective on the product's development. It is common to invite technical specialists from outside the team to help resolve design issues.

Variances from plan are elevated to the chief engineer, who either approves the variance or requires that an appropriate corrective action be taken."

> **4.4.4 Design Input**
>
> Design-input requirements relating to the product, including applicable statutory and regulatory requirements, are to be discussed and contract review activities are to be included.

"Design Input

The program manager and project engineer are responsible for the functional specifications based on various internal department and customer inputs and on Growth's decision of the product definition.

The functional specifications include all statutory and regulatory requirements, for example, use of the CE mark, and require the final approval of the chief engineer. The product proposal results from this specification, and its scope and complexity are proportional to the program's cost. Any ambiguities or conflicts that result from the functional specification are brought before the chief engineer for resolution."

> **4.4.5 Design Output**
>
> Specifications are to be documented and expressed in terms that can be verified against the design-input requirements and validated.

"Design Output

The chief engineer is required to ensure that the final functional specification agrees in detail with the customer's requirements, and only then approves and releases the document.

Upon release of the final functional specification, the design team develops the User's Manual, which contains instructions for any special handling procedures that may be required—for example, ESD handling instructions—and product maintenance.

In addition, diagnostic test procedures are developed by the project engineer to verify and validate the product prior production hand-off."

> **4.4.6 Design Review**
>
> At appropriate stages of design, formal documented reviews of the design are to be planned, conducted, and recorded.

"Design Review

The program manager schedules design reviews into the program plan based on the complexity of the program, and keeps the minutes from these reviews in the program files.

Several types of reviews are scheduled, including technical reviews to define design, code, algorithms, and production and manufacturability; and marketing and sales reviews to ensure that the program will meet the customer's changing needs in a dynamic marketplace.

Appropriate guests are added to the design team as required, to cover specific technical and/or marketing issues."

> **4.4.7 Design Verification**
>
> At appropriate stages of design, design verification is to be performed and documented.

"Design Verification

All hardware products are verified before release into production by means of a final prototype build and test, to ensure that all supporting documentation for production is available and correct.

All software products are verified before release into production by means of a verification copy of the software to ensure completeness of the production transfer package.

The program manager and project engineer are jointly responsible for performing and documenting the verification process and its results and the completeness of the program files in this regard."

> **4.4.8 Design Validation**
>
> Design validation is to be performed to ensure that the product conforms to defined user needs and/or requirements.

"Design Validation

All hardware and software products are validated before release into production to ensure conformance to the final functional specification—which includes all customer requirements.

This process requires a system-level test strategy using a typical customer's system. In certain situations, the customer may ask to be present and validate acceptance through his or her signature.

The responsibility for performing and documenting the validation testing is jointly held by the program manager and the project engineer, who are also required to include a complete record of the activities in the program files.

A continuing engineer moves with the project from pilot line runs into forward production, and then remains with the program until manufacturing engineering phases in. An ECO is used for this transfer."

4.4.9 DESIGN CHANGES

All design changes and modifications are to be controlled effectively and recorded.

"Design Changes

Design changes that result in variations from the functional specification are reviewed by the chief engineer for approval. The change, or rejection of the change, is documented in the program files by the program manager.

The chief engineer has the discretionary authority to call a more general management review if required.

An engineering change order (ECO) is used to release products to production, and once the product is released any changes are made via the ECO process."

DOCUMENT AND DATA CONTROL POLICY

Response

Paul and Charlie made the decision to do a joint first draft and then call the team together to critique and edit the document.

The process document, entitled "Document & Data Control Processes," turned out to need considerable revision as the group agonized over just how documents were to be distributed and maintained. They quickly realized why this was the area of highest nonconformances in the actual certification audit. Some of their ideas that had appeared to be logical broke down when they tried some test passes at revision control.

It took a week, off and on, to arrive at a complete first draft, and that work is shown in "Section 5."

4.5.1 GENERAL

Growth is to establish and maintain documented procedures to control all documents and data..

"Section 5 Document and Data Control Policy Procedure

The document that describes Growth's approach to the control of all hardware and software documentation and data is entitled "Document & Data Control Processes."

Externally received documents—for example, ISO 9000 standards, vendor documents, and customer specifications—are controlled locally by the appropriate user. Such documents are used in either the design, manufacture, or preventive maintenance of product and associated instrumentation."

4.5.2 Document and Data Approval and Issue

The documents and data are to be reviewed and approved for adequacy by authorized personnel prior to issue, and are to be revision controlled in regard to distribution and obsolescence.

"Control

All released engineering documents are under ECO control whether in hard copy or on-line. Program managers and project engineers have ready access to such documents via the engineering computer system (ECS).

All policy, process, and procedural documents are controlled by means of the code numbers and "red" control numerals described in Section 2 of this Manual. Each document is assigned an "owner" who is responsible for the review, approval (sign-off), and release of the documents.

A master document list is maintained by both the quality documentation supervisor and the engineering documentation manager to ensure correct distribution and that users have the most recently revised documents at their work sites. Tier 4 documents, for example, forms contained in a Forms Master Manual, are controlled by the quality documentation supervisor and distributed to the local managers upon request.

Memos, reports, and similar documentation are controlled at the local manager level and do not require a master list.

Obsolete documents are removed from use at the local manager level upon receipt of revised documents.

Documents that are maintained for legal and informational purposes are either marked accordingly on their containers and archived and maintained by the accounting department, or by the chief engineer, as appropriate. Department managers are authorized to determine which obsolete documents are to be retained."

4.5.3 Document and Data Changes

Changes to documents and data are also to be controlled in the same effective manner as the original issues, and changes are to be recorded, as appropriate.

"Revision Control

Changes at the policy, process, procedural, and forms level are made via the department change order (DCO). Changes to released hardware and software engineering documents are incorporated via the engineering change order (ECO).

Document and data change orders for hardware and software products are reviewed and approved by the designees defined in the process, who are usually the document owners or their designees.

Previous revisions and pertinent background information are directly available as part of the DCO and ECO formats. Such formats also contain descriptive material for the nature of the change.

For those documents on-line, the systems are periodically backed up and stored electronically by the MIS Manager."

PURCHASING POLICY

Response

Shelby decided to do a first draft on her own since the process documents that described the overall purchasing protocols for Growth had been extremely complete. Such documents had given her confidence in her ability to summarize the entire program in a relatively short time. She would then hold a general review with the team for editing purposes.

She decided to place all of the documents related to purchasing in a controlled three-ring binder, which she aptly named the "Purchasing Manual."

The move proved to be popular with her team, and the entire first draft was completed within a week. We show her work in "Section 6."

4.6.1 GENERAL

Growth is to establish and maintain documented procedures.

"Section 6 Purchasing Policy Procedures

The purchasing documents include a set of procedures and specifications required when procuring subcontractor services, piece parts, noninventory, and other items. This information is contained in a controlled three-ring binder entitled the "Purchasing Manual" and maintained by the purchasing manager."

4.6.2 EVALUATION OF SUBCONTRACTORS

Growth is to effectively manage subcontractors and maintain records of approved subcontractors.

"Evaluation of Subcontractors

Growth initially selects its subcontractors on the basis of site visits, questionnaires, quotations, and references. In all cases, a long-term commitment to quality assurance, especially in terms of ISO 9000, is sought with its key customers. After selection, site quality audits, when appropriate, are conducted jointly by the quality assurance department and manufacturing department to maintain an ongoing partnership relationship. In addition, periodic evaluation reports, which inform vendors of their progress against on-time delivery, performance, and quality are published by the purchasing manager.

PCB design houses are specifically evaluated by the director of manufacturing and the director, quality assurance. Certificates of compliance and/or analysis are required from the PCB design house. The purchase of noncritical components is entirely at the discretion of the buyers.

The records of approved subcontractors are maintained in the approved vendor list (AVL) maintained by the purchasing manager. Subcontractors are either added or removed from this list through a continual evaluation process documented by the purchasing manager."

4.6.3 Purchasing Data

Purchasing documents are to *contain data* clearly describing the product ordered. The purchasing documents are to be reviewed for adequacy of the specified requirements prior to release.

"Purchasing Data

Growth's purchase orders (P.O.s) include the P.O. number, the date the order was placed, the P.O. type, the date the P.O. was last changed, the quantity ordered, the part number and its description, the vendor, the price, and the required quality standards, when appropriate. The quality requirements are supported by attached engineering drawings and specifications.

The P.O.s are controlled in a numbering sequence that is kept in logs controlled by the buyers. Two different types of P.O.s are used to purchase either inventory or noninventory items.

All inventory P.O.s require the signature of the purchasing manager prior to release. Noninventory orders are signed off by the buyers."

4.6.4.1 Supplier Verification at Subcontractor's Premises

Growth is to effectively manage source inspection, if required.

"Supplier Verification at Subcontractors

Growth performs source inspection, when appropriate, at the subcontractor's facility. In those situations, the purchasing manager is required to formally alert the vendor prior to the visit and develop a protocol that establishes the conditions under which the source inspection can be terminated and future shipments can be released."

4.6.4.2 Customer Verification of Subcontracted Product

Growth is to permit its customers to source inspect at its subcontractor's premises and at Growth's facilities.

"Customer Verification of Subcontracted Product

If agreed to contractually, Growth allows its customers—or their representatives—to inspect their product at either Growth or at the subcontractor's facility.

Growth neither uses the positive results of such an inspection as evidence of its effective vendor management, nor feels that it is a means of releasing Growth from always supplying acceptable product to its customers nor that it affects the potential rejection of such product by the customer."

CONTROL OF CUSTOMER-SUPPLIED PRODUCT POLICY

Response

Gary called his small but intense group together to create the policy for customer-supplied material. The entire effort took a few hours and was confined to essentially the returned materials authorization (RMA) procedure. Their work is shown in "Section 7."

> **4.7 CONTROL OF CUSTOMER-SUPPLIED PRODUCT**
>
> Growth is to establish and maintain documented procedures and control customer-supplied material and product as if it were their own.

"Section 7 Control of Customer-Supplied Product Policy Protocols

Customer-owned (supplied) materials at Growth consist of RMAs and miscellaneous engineering test systems and development equipment used in the design validation process.

The RMAs are controlled by the RMA procedure and tracked by customer service, which has the primary responsibility to respond effectively to customer returns.

Customer service also provides the required management for field exchanges, loaners, or evaluation units—which are handled in the same manner as product returned for repair or refurbishment.

Miscellaneous test systems and development equipment is maintained and tracked by local area engineering managers.

Any customer-owned equipment or material that is lost or damaged in any way is reported to the customer service manager for disposition and corrective action with the customer."

PRODUCT IDENTIFICATION AND TRACEABILITY POLICY

Response

Karen actually worked elements 4.8, 4.9, and 4.12 at the same time with her stewardship group. What we will show you is how the compositions looked in the Manual as they came up in the natural ISO 9001 sequence.

In spite of a fairly complex set of elements, the process document entitled "Hardware & Software Manufacturing Processes" had set the stage completely for this summarization, and so the group moved forward very quickly and completed all three drafts in about four hours.

> **4.8 PRODUCT ID & TRACEABILITY**
>
> Where appropriate, Growth is to establish and maintain documented procedures for identifying and recording the product by suitable means throughout the entire product life cycle.

"Section 8 Product Identification and Traceability Policy Product Identification and Traceability

Growth uses product identifiers (P.I.s) to define the sales & marketing product name, which may not be the same as the part number. The director of sales & marketing maintains a list of the approved P.I.s. All hardware and software documentation for a given product clearly states the P.I.

The director of manufacturing is responsible for determining the serial numbers of all hardware and software products, and all products are labeled with either the P.I. or the part number and its revision level. All hardware products are labeled with their serial number, and all software products are labeled with their release date. Software products are identified and traceable by part number and revision level.

The director of manufacturing maintains a data base that tracks board history.

Hardware documentation includes information needed to trace and record all printed wiring assemblies, box, and system-level products by board type and manufacturing lot. This information includes BOM, build, and test information.

An MRP system is used to identify and track the progress of units as the lot moves through the manufacturing process. All product is followed and tracked by job number.

Job process tags are attached to all product for use in identifying, tracking, and tracing the product through the factory. Each operation in the process is verified by the operator's initials. Different colors are used to differentiate standard product from returned product."

Process Control Policy

Response

Karen and her group continued their work onto element 4.9 and we see that work in "Section 9."

> **4.9 Process Control**
>
> Growth is to identify and plan the production, installation, and servicing processes that directly affect quality and ensure that these processes are carried out under controlled conditions.

"Section 9 Process Control Policy The Manufacturing Process

Growth's manufacturing process is based on ten stages, as graphically demonstrated in Case Study Figure 5 (refer to the process document entitled "Hardware & Software Manufacturing Processes"):

Planning

The director of manufacturing oversees the functions of production and materials control. The production control manager and the materials manager analyze the sales forecast and inventory status and produce production schedules that are approved by the director. The schedules drive the procurement process.

Floor documentation released by the control team includes a process-control document, a color-coded assembly drawing, required inspection visual aids, and special mechanical drawings as required.

Monitoring and Control

Growth uses a series of software-based tests that display various pixel image patterns for visual definition, and test procedures are developed by design engineering for specific product lines.

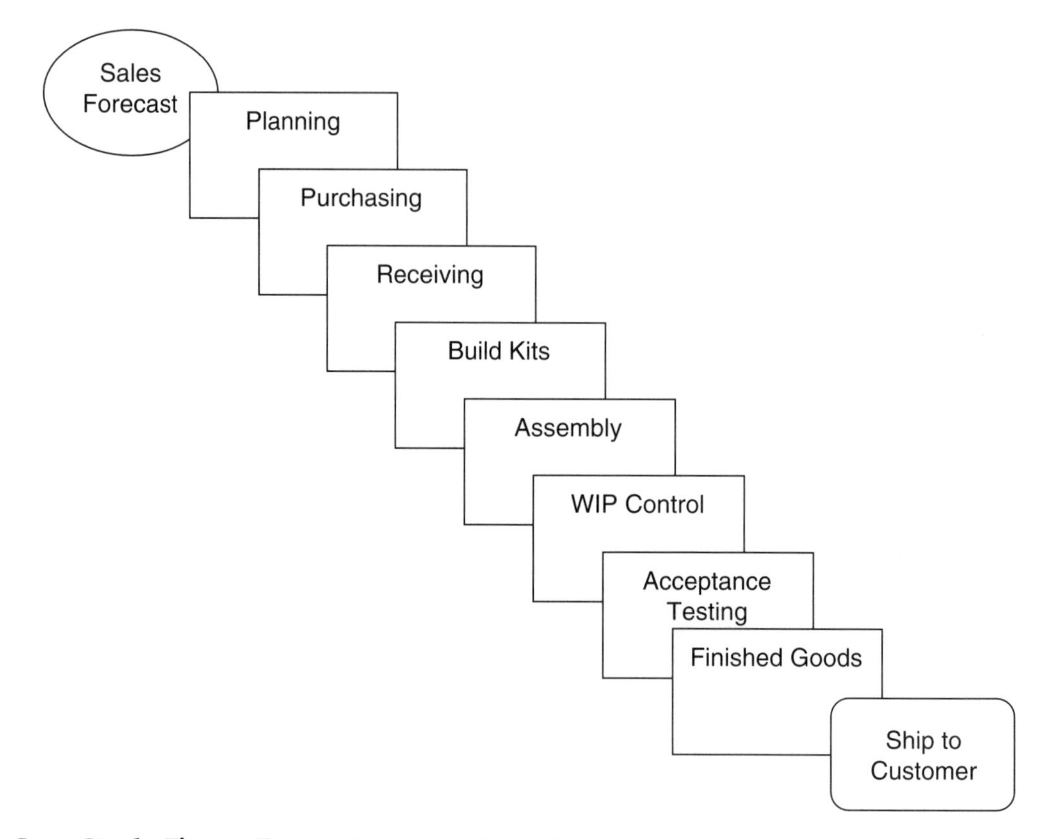

Case Study Figure 5 Product manufacturing protocols.

The controlling process document is the "process control document (PCD)" issued by engineering services for each assembly built. This document contains both customer documentation and Growth's process revision control system to ensure that the current product is built to the correct revision. This document fully defines each process step that an assembly must undergo. These documents are prominently displayed in the materials, surface mount, through-hole, quality control, and other areas as required when an assembly is in process.

This document also identifies all required tooling, programming, and special instructions as necessary. In addition, any component substitutions or special material preparation is included.

The document requires sign-off by materials, process engineering, manufacturing, quality, and when appropriate test supervision. The revision levels are controlled and updated under the ECO process.

Each process step is monitored and, when applicable, statistical techniques are used to measure process variables to ensure that the process is in control. Customer quality requirements have absolute precedence.

Work Flow

The layout of the process area is designed to allow for an efficient and flexible process flow. Duplication of effort and retracing is minimized.

New Processes

The introduction of either new processes or new instrumentation into production requires the approval of the director of manufacturing as controlled by the ECO process.

Workmanship

Functional test suites are run against all hardware products to ensure that the products are functionally correct. Video quality is verified by running a series of software tests. IPC standards are used where applicable in the assembly process.

ESD Control

ESD control is required throughout the manufacturing process, and the implementation and training in ESD is the responsibility of the reliability engineering supervisor.

Preventive Maintenance

A full program of preventive maintenance ensures long-term reliable operation of the capital equipment. Maintenance is performed at varying frequencies, as required, and is the responsibility of the maintenance engineer.

Environmental Controls

The work environment is maintained to ensure that the processes are stable and employees are physically comfortable. For this purpose, four automatic temperature control thermostats are located throughout the manufacturing area.

Nonrepetitive Processes

Nonrepetitive processes, such as prototype builds, depend upon very close customer interfacing to define levels of process controls. Special instruction sheets are issued as part of the process control document, when applicable."

> Growth is to discuss its use of processes requiring prequalification of their process capability, that is, special processes. Records are to be retained for such processes.

"Special Processes

Special production processes, such as conformal coatings, are closely monitored and conform to strict qualifying procedures as defined by customer requirements. Records are maintained by the director of manufacturing related to the qualified process used, the qualifications of the personnel, and the qualification of specific equipment used.

In addition, because image quality is measurable on both a quantitative and subjective level, acceptance testing of product is performed by qualified test operators using calibrated test equipment. As with the special production processes, pertinent records are maintained by the director of manufacturing for this purpose."

INSPECTION AND TESTING POLICY

Response

Mike decided to compose the inspection and testing policy on his own and then review it with the group. He felt confident in this task because of the hard work that had gone into the process document entitled "Inspection & Testing Processes." It seemed like a fairly straightforward task to transform the process into a series of policy quality statements. It was.

The entire process with the group took no more than half a day. We see their work in "Section 10."

4.10.1 GENERAL

Growth is to establish and maintain documented procedures and records.

"Section 10 Inspection and Testing Policy Procedure

Inspection and testing protocols at Growth are created, controlled, and recorded by quality assurance and include reliability testing at accelerated temperatures over extended test times.

The director of quality assurance is responsible for all inspection and test procedures at Growth. The primary document for this activity is contained in "Inspection & Testing Processes."

4.10.2.1 GENERAL

Growth is to ensure that incoming product is not used or processed until it has been adequately accepted.

"Incoming

All raw material received from vendors is verified by receiving and sampled by quality control to ensure conformance to specification. Product that does mot meet specification is placed in MRB for disposition. In addition, printed circuit boards require a certificate of conformance by the contract vendor with deliveries.

In cases of immediate or urgent production requirements, a waiver procedure is used to identify material released for production prior to the issuance of a complete documentation package."

4.10.3 In-process Inspection and Testing

Growth is to inspect and test the product as required during processing.

"In-Process

Each key step of in-process inspection and testing is monitored by quality control personnel.

Specific check points include

First Article: A first article inspection process is performed on the first board of every production run. A distinct work tag is attached to identify that assembly. Authorization to commence the production run is given when the board passes this procedure.

Pre-Reflow: A pre-reflow solder inspection is performed on 100 percent of boards requiring the reflow soldering process. This inspection is performed immediately after the pick-and-place process by manufacturing personnel.

Component Side: A top-side, in-process inspection is performed on 100 percent of boards built on either the prototype or production lines.

Pre-wave: A pre-wave solder in-process inspection is performed on 100 percent of the boards requiring the wave solder process.

Bottom-Side: When applicable, a bottom-side inspection is performed on boards built on either the prototype or production lines. This inspection is performed by quality control personnel.

Special In-Process Tests: Various types of electrical, functional, or visual testing are performed on a board at particular process locations, as required. This testing either conforms to customer requirements or is deemed appropriate to ensure conformity to customer or Growth quality standards."

4.10.4 Final Inspection and Testing

Growth is to carry out final inspection and testing and document evidence of acceptance.

"A final inspection is performed on 100 percent of boards built on either the prototype or production lines. All final inspections are performed by quality control personnel.

Finally assembled products are visually inspected by operators to verify that the products have the appropriate stamps or tags. Product that undergoes functional tests and burn-in is logged into the first-pass yield data base for analysis. All inspected and tested product receives a stamp to indicate completeness."

4.10.5 Inspection and Test Records

Growth is to establish and maintain records that provide evidence that the product has been inspected and/or tested. Pass/fail data are to be retained.

"Records and Nonconformances

Any material or assembly found in nonconformance is promptly identified and segregated as such. Prompt action is then taken to bring the material back into conformance or for disposition according to the protocols for the control of nonconforming product, as discussed in Section 13 of this Manual.

The records of acceptance and/or rejection, which clearly show the responsible inspection authority, are maintained in their respective areas, that is, in either materials or quality assurance files.

Notification of products on hold is identified by a QC verbal communication to all affected areas. Any changes required to remove the product from hold are done through the ECO process.

All product that is shipped from Growth meets all required specifications. The shipper is required to verify the inclusion of all product and related components into the shipped package, including software and documentation as noted on the pick ticket for the order. The shipper validates the pick ticket with date shipped and his or her initials.

Software

For required software, the software files to be replicated are revision controlled by means of the ECO process and its associated part number. Error checking is built into the replication process.

Records

All records are maintained by either quality assurance or materials, as appropriate."

Control of Inspection, Measuring, and Test Equipment Policy

Response

Due to the complexity of the subject, Mike decided to call the group together and do a joint first draft. They began with lists of equipment and agonized over the decisions on which equipment really needed to be calibrated. The effort took about two weeks overall to complete, but they felt satisfied that the facility had been adequately covered.

Their decision on calibration required that only equipment that was either used for acceptance or rejection of product, or was used to make absolute measurements—for example, in the creation of an engineering specification—required calibration. No other IM&TE was to be labeled if calibration was not required unless the customer demanded it be done ("No Calibration Required"). In this manner, they estimated their annual savings at over $5,000—all bottom-line dollars.

Their work is shown in "Section 11."

> **4.11.1 General**
>
> Growth is to establish and maintain documented procedures to control, calibrate, and maintain inspection, measuring, and test equipment (including test software). This includes measurement devices.

"Section 11 Control of Inspection, Measuring, and Test Equipment Policy Calibration

Growth requires the calibration and maintenance of equipment used to either make absolute measurements or accept or reject product—for example, oscilloscopes, multimeters, photometers, and temperature probes—according to

Growth's "Metrology Manual." Only IM&TE that requires calibration is so marked.

Testing/Software

Test protocols are initially designed by design engineering. They are part of the release package to manufacturing, where they are maintained under ECO control.

Testing/Hardware

Test hardware used by Growth for development and production is maintained and/or calibrated by design engineering.

Measurement Uncertainty

Calibrated equipment has documented tolerances. Where applicable, measurement uncertainty is determined by design engineering.

Records

Calibration information is maintained by quality assurance and includes equipment type, unique identification, frequency of calibration, calibration method, acceptance criteria, and actions required if equipment becomes uncalibrated. This also includes any calibration and/or maintenance falling outside the specified calibration intervals.

Technical Data

If it is contractually required, Growth makes available to a customer all technical data related to the specific measurement equipment."

4.11.2 CONTROL PROCEDURE

Growth is to effectively control, monitor, and record this process. The metrological confirmation system for measuring equipment given in ISO 10012-1:1992 may be used for guidance.

"Control Policy

Responsibility

Either the equipment manufacturer or design engineering is responsible for determining the accuracy and precision of any purchased testing or measurement system. The required accuracy and precision of the equipment is established as either a part of the selection process or when otherwise specified by the customer.

Calibration

Calibrated equipment is visibly tagged or labeled, indicating the last calibration date and expiration date. This label also has the authorized signature of the person who performed the calibration.

Calibration information is maintained by the director of quality assurance. It includes the definition of nationally and/or internationally recognized standards, equipment type, unique identification, frequency of calibration, calibration method, acceptance criteria, and actions required if equipment becomes uncalibrated. This also includes any calibration and/or maintenance falling outside the specified calibration intervals.

Standards

All calibrated equipment is calibrated against international or national standards, as appropriate. In some cases, internally created test protocols are used as test standards based on actual applications. Calibration plans are managed via "logs" that are maintained by quality assurance to indicate calibration cycles and frequency.

Calibration labels are used on all required IM&TE to alert operators that calibration is adequate or due. If calibration is overdue, operators are to immediately alert quality assurance and suspend use of the equipment until calibration is completed.

All equipment is sent out for independent calibration to companies selected based on their capability of using appropriately known standards. Quality assurance maintains logs of all of these transactions. A Paradox database file, CALIBRAT.DB, is maintained listing calibration status for all equipment on a calibration cycle.

Invalidation

Any product that has been validated with uncalibrated equipment is subject to a documented joint review by quality assurance and the director of manufacturing. The CAR protocols are used when required.

However, it is the responsibility of each operator to check equipment calibration status prior to each measurement. Appropriate actions are taken to correct any situation in which measurements were made with equipment found later to be out of calibration. Such actions include notifying the customer, retesting product, product recall, waivers, and/or rework.

Conditions

All Inspection Measuring & Testing Equipment (IM&TE) requires room-temperature operation, only, and no special handling other than normal maintenance as prescribed in the equipment's operation manual.

Safeguarding

Calibration labels and/or seals are placed in an appropriate location to prevent adjustments. If the label/seal is broken, the calibration becomes invalid and the equipment may not be used until recalibrated. Operators are not allowed to make any adjustments to equipment. All adjustments are under the control of quality assurance.

Subcontractors

IM&TE used by Growth's assembly houses, when required, is consigned to the house and maintained using Growth's calibration policies."

INSPECTION AND TEST STATUS POLICY

Response

Karen and her group continued their work from element 4.9 to element 4.12, and we see that work in "Section 12."

> **4.12 INSPECTION & TEST STATUS**
>
> The inspection and test status of product is to be **identified** by **suitable** means.

"Section 12 Inspection and Test Status Policy Procedure

The inspection and test status process is contained in the "Hardware & Software Manufacturing Processes" document.

When Growth receives components, printed circuit boards, and assembled materials, the receiver verifies and dates the packing slip. Upon receipt, quality assurance verifies the assembled material's inspection and test status.

Job Tags

Job tags that define job numbers and individual board serial numbers are used to monitor manufacturing and inspection process traceability on all boards processed. All manufacturing and test processes are initialed and dated on the tag by the operator performing the specific process. All inspection processes are initialed and dated on the tag by the inspector performing the specific inspection.

Inspection Stamps

Inspection status is further supplemented with the use of unique inspection stamp markings directly onto the surface of the printed circuit boards. The stamps use an indelible ink that was selected to withstand all cleaning processes.

Either inspection and test stamps are marked on the assemblies as required before shipment. The status of software is determined by the presence of the label that includes the appropriate part number and revision.

Records

Growth's in-house quality test records that define the hardware status are maintained in the first-pass yield database.
"Overview." This database tracks boards by the product name, build lot, and serial numbers. The "Growth RMAs" database is used to record any nonconformance of both hardware and software products.

Location

Physical locations in the production area for work in process and finished goods are also used for identifying the status of assembled product.

Nonconformance

Rejected or failed material is marked with a nonconformance color coded tag, specific to the type of nonconformance.

Conformance and Concession

Products are required to be fully tested and burned-in prior to shipment. If a product that does not meet full specification is shipped, a mutual agreement is established between sales and the customer.

Notification

A "hold" tag is used to identify any product that may be placed on hold for any reason. Any subsequent changes to product are identified via the ECO Process."

CONTROL OF NONCONFORMING PRODUCT POLICY

Response

Mike felt that this particular element could be done first by himself and then with a review by the group. The document entitled "Control of Nonconforming Process" had been quite complete and had spelled out the MRB process in such detail that he was able to do the first cut in just a few hours. The group had some minor comments, but they felt the policy was right on.

> **4.13.1 GENERAL**
>
> Growth is to establish and maintain documented procedures to ensure that product that does **not** conform to specified requirements is prevented from unintended use or installation.

"Section 13 Control of Nonconforming Product Policy General Process

The process by which Growth identifies, documents, evaluates, separates, and disposes of nonconforming product, and notifies the appropriate executive functions, is contained within the document entitled "Control of Nonconforming Process."

A material review board (MRB) is used to make nonconforming product and material decisions. The group is represented by quality assurance, manufacturing, and continuing engineering, and the MRB Report is used to notify the affected areas."

4.13.2 Review and Disposition of Nonconforming Product

The responsibility for review and authority for the disposition of nonconforming product is to be defined. Repaired and/or reworked product is to be reinspected.

"MRB Options

Dispositions include rework, acceptance with or without repair by concession, and swap/replacement.

Concessions

Concessions are achieved through direct contact between Growth's sales & marketing department and the customer.

Rework

If a product fails in test and can be reworked, it is retested and reinspected. Product that cannot be reworked is tagged and segregated for further disposition by the MRB. For example, rejected products are identified and segregated from conforming product into one of three areas—the rework area on the manufacturing floor, the WIP shortage rack on the manufacturing floor, or the nonconformance storage cabinet in the stockroom.

Failed Parts

Other products, such as SIMM modules, cable assemblies, and components that fail in the process or in the field are dispositioned by segregating the material into an MRB location. Purchasing notifies the appropriate vendor for a return authorization number and returns the material to the vendor.

Records

If agreed contractually, where nonconforming product is used as a result of concession with either the customer or the customer's representative, the nature of the nonconformity and the method of repair is recorded and reported to the customer. Sales & marketing handles this interface.

Reinspection

In all cases, repaired and or reworked product is reinspected prior to shipment."

Corrective and Preventive Action w/Customer Complaints Policy

Response

For the 4.14 element, Mike wanted the whole team on board all the way and worked out the first draft in a morning's workout in the main conference room. The process document, "Corrective & Preventive Action Processes w/Customer Complaints," had been well done, but there were quite a few open questions in the preventive action area that only the group could straighten out.

In fact, it was some time before there was agreement that preventive actions would count only if they were indeed proactive and affected the overall operation, not just one particular product. Otherwise, they were listed as corrective. It seemed a small point at first, but it turned out to be a key ingredient in Growth's eventual business successes.

Sam had also indicated that customer complaints was to be a major section in this policy and reinforced the idea when he presented the registrar's directive in this area. This element was just too important for the success of the program not to have everyone totally aware of all points in the activity. Sam was concerned enough to publish minutes of the meeting.

The group was satisfied that they had caught the spirit of the element, and we present their composition in "Section 14."

4.14.1 General

Growth is to establish and maintain documented procedures for implementing corrective and preventive action.

"Section 14 Corrective and Preventive Action w/Customer Complaints Policy Procedure

The document describing corrective and preventive action is entitled "Corrective & Preventive Action Processes w/Customer Complaints." This process covers the following programs:

- Audit program

- Customer complaints
- Customer returns
- Product issue reports that include software and hardware issues, and product feature requests
- Customer/field reports
- Engineering and documentation change control
- First-pass yield and quality data overview

These areas form the data input for the corrective and preventative action program.

Responsibility

The management of this program and the analysis of these data are the joint responsibility of the directors. The chair of the group is the director of quality assurance.

Level of Action

The level of corrective and/or preventive action taken by Growth depends upon their degree of impact on the product lines. These decisions are made by the directors, who can also assign decision making to local area managers, as required. Another way of resolving specific critical issues is by the formation of an engineering and support crash team, when needed, under the management of the director of design engineering.

Revisions

The corrective and preventive action program uses the document and data control process to provide a means to make any documentation change required as a result of this process."

> **4.14.2 Corrective Action**
>
> The procedures for corrective action are to include the effective handling of customer complaints and reports of product nonconformities and recorded investigations.

"Corrective Action

In-Plant Operations

In-plant operations include all in-plant inspection and acceptance testing as well as the RMA system. All nonconformances are identified, corrected, and verified at Growth.

These date are is contained in databases managed by quality assurance, who review and analyze the information. The director of quality assurance reports both these findings and the status of subsequent corrective action requests to top management in the management review meetings.

Handling CARs

When a corrective action request is presented to a local area manager, it is the local area manager's responsibility to take timely action in defining and eliminating the root cause of the nonconformance.

Verification

The presenter of the CAR is responsible for ensuring that the corrective action is taken and that it was effective, so that closure can occur.

Audit Program

The director of quality assurance is responsible for managing the total quality audit program. This program includes internal audits, SCARS, third-party audits, and customer audits. The status of this program is presented at the management review meetings.

Subcontractor Verification

Corrective actions that are a consequence of subconntractor evaluation—either from results of incoming inspection, such as return to vendor items (RTV) or final test, or from on-site inspection—are managed by the director of quality assurance with the assistance of the purchasing manager.

Product Development Verification

Corrective actions that result from engineering activities, which include the hardware and software databases, are the responsibility of the director of design engineering.

Managers

All managers are responsible for the detection, analysis, and eventual elimination of potential causes of nonconformities through the examination of available data. There data include returned material, customer complaints, discrepant material reports, design review, and current documentation.

Plan

All managers are responsible for generating the plan to remove potential causes of the nonconformity, and to ensure the plan results in the effective control of such actions.

Action

All managers are responsible for collection and analysis of data within their respective areas. From the analysis of these data, the managers are responsible for deciding the appropriate action to be taken, which can include formation of quality improvement teams, assignment of tasks, and issuing ECOS.

Customer Complaints

Responsibility

The customer service manager is responsible for collecting, analyzing, and generating corrective actions related to customer complaints.

The customer complaints are obtained through the following sources:

- Direct customer contact via phone, fax, or E-mail
- Field sales representatives or distributors

Resolution

The corrective and preventive action programs are used to resolve customer complaints. Resolution is generally through the "Customer Service Complaints Logs" and the "SCAR/CAR" databases.

Any form of customer feedback that is received can result in a CAR. As a result, this policy is designed to service the needs of the customer, yet minimize the burden placed upon the customer due to procedural requirements.

Status

The director of sales & marketing presents the status of customer complaints at the management review meetings."

> **4.14.3 Preventive Action**
>
> The procedures for preventive action are to include the use of appropriate research sources, the determination of the steps needed to deal with relevant problems, the initiation of preventative action and application of controls to ensure effectiveness, and review by top management.

"Preventive Action

Reporting

The director of quality assurance coordinates the reporting at the management review on the status of any action plans that are taken in the area of preventive action.

For this purpose, data are analyzed from manufacturing, sales & marketing, quality assurance, and design engineering to detect, and eliminate potential causes of nonconformities. Such data are displayed as Pareto charts on a monthly basis and reported as part of the management review process. A list of preventive actions is maintained to indicate progress in this important area.

Preventive actions are formed from all administrative and operational areas—improvements in the use of people, machines, instrumentation, facilities, and test procedures."

HANDLING, STORAGE, PACKAGING, PRESERVATION, AND DELIVERY POLICY

Response

There was no question in Karen's mind in regard to the first draft. She would do it and review it with the team. The activities that had already occurred were all she needed for this section. She was right.

> **4.15.1 GENERAL**
>
> Growth is to establish and maintain documented procedures.

"Section 15 Handling, Storage, Packaging, Preservation, and Delivery Policy General Process

Growth's process for handling, storing, packaging, preserving, and delivering product is contained in the "Hardware & Software Manufacturing Processes" document."

> **4.15.2 HANDLING**
>
> Growth is to provide methods of handling product that prevent damage or deterioration.

"Handling

All Growth employees are trained to follow Growth's ESD practices and to know the effects ESD can have on Growth's products.

Quality assurance is responsible for the maintenance of the ESD equipment and documentation. The adherence to these policies is the responsibility of the department managers and all employees."

4.15.3 Storage

Growth is to demonstrate effective storage procedures.

"Responsibility

Receiving and quality assurance receive and approve all product that is shipped into the facility. Once the products are verified and received, they are put into a secured stockroom for storage. The stockroom supervisor is responsible for handling the transactions and activity in and out of the stockroom to prevent any damage and/or misplacement of components.

Signout

Parts and finished goods leaving the stockroom area must be noted out on a sign-out worksheet. The stockroom is organized by Growth's part numbers. The inventory control supervisor monitors these components on a regular basis."

4.15.4 Packaging

Growth is to control packing, packaging, and marking processes.

"Packaging Procedure

Growth has a specific shipping procedure that must be followed when shipping customer products. The manufacturing process directs the shipper to the necessary documentation and equipment.

Packaging Material

The manufacturing processes describe the specific packing materials that are used. These materials include static shielding bags for all board-level product. When requested, Growth packages product according to special customer specifications.

Supplies

The materials being used for packing are monitored by the shipper/receiver and the appropriate buyer."

4.15.5 Preservation

Growth is to apply appropriate methods for preservation and segregation of product.

"Preservation

There are separate areas designated for finished goods, WIP, raw components, customer-owned material, and MRB. These areas are closely monitored by quality assurance. Growth adheres to shelf-life requirements where applicable."

4.15.6 Delivery

Growth is to arrange for the protection of its products after final inspection and test, and where contractually specified, this protection is to extend to include delivery to destination.

"Delivery Policy

Growth has a segregated secured stock room for finished products that have passed final test and inspection. These products are available to ship for customer orders. Products are shipped FOB, the Growth Division, St. Louis, MO. When the customer does not specify a delivery method, Growth uses the most expedient, cost-effective, and quality assured method.

Delivery to Destination

If there is a contractual agreement, Growth extends its protection responsibility to include delivery to destination."

Control of Quality Records Policy

Response

Paul brought the team together to deal with quality records. This was an area that never seemed to end. Indeed, they were adding records to the master list an hour before the assessors showed up for the certification audit.

The policy, however, went together rather easily based on the process document entitled, "Quality Records Procedure & Master List." The group was quite satisfied with Paul's draft—which we show in Section 16.

4.16 Control of Quality Records

Growth is to establish and maintain documented procedures to effectively control its quality records, including recorded retention times. Where agreed contractually, quality records are to be made available for evaluation by the customer for an agreed period.

"Section 16 Control of Quality Records Policy Procedure

The "Quality Records Procedure & Master List" details the procedure for identifying, collecting, indexing, accessing, filing, storing, maintaining, and disposing of quality records. Maintenance of quality records is the responsibility of each local area manager.

Quality records are identified by either title or number. The "Quality Records Procedure & Master List" indicates the location of the records. They may be on-line or hard copy. The master list includes who is responsible for the record, its retention time, and its status (on-line, hard copy, obsolete, or obsolete but retained).

Subcontractor data, in the form of certificates of compliance or analysis, are included in the master list.

Filing

The hard copy records are filed in either commercial-grade steelcase files or in corrugated containers such as banker's boxes. On-line records are managed by the MIS manager. All files are kept on-site. Records are kept within the easy reach of the users to facilitate usage.

Legibility

The quality records are typed for legibility whenever possible. Handwritten records are written with permanent-black-ink pens.

Conformance

Growth uses its quality records as a key source of information for presenting the status of its internal audit, and for a preventive and corrective action program provided to management for review. Most importantly, the records are a source of quantitative information, used in trend analysis.

Process Control

Many records are kept by the local area manager, for example, process control records are kept in the individual customer job folders in production and include, when available, assembly drawings, bills of materials, fabrication drawings, visual aids, process control documents, and measurement documents.

Contractual

In those cases when Growth enters into a contractual agreement, records are made available to the customer or its representative for evaluation for a limited period."

INTERNAL QUALITY AUDITS POLICY

Response

Mike knew that the extensive work the group had already done on the defining document, "Quality Audit Processes," dictated that he do the first draft and then review it with the team. The draft took only two hours, and the group approved it is less than an hour. It looked solid.

> **4.17 INTERNAL QUALITY AUDITS**
>
> Growth is to establish and maintain documented procedures for planning and implementing internal quality audits. Guidance on quality-system audits is given in ISO 10011-1: 1990: Guidelines for auditing quality systems—Part 1: Auditing; and ISO 10011-2 : 1991: Guidelines for auditing quality systems—Part 2: Qualification criteria for quality systems auditors; and ISO 10011-3:1991: Guidelines for auditing quality systems—Part 3: Management of audit programmes.

"Section 17 Total Quality Audits Policy Procedures

Growth's procedure for internal quality audits is entitled "Quality Audit Processes." All audit protocols are managed by the director of quality assurance.

The document details the

- Selection, training, and proficiency requirements of internal quality auditors
- Way in which auditor independence is ensured by the controller
- Gathering of specific information concerning the area to be audited by the auditors
- Use of the auditor's checklist, which is an integrated ISO 9001 and ISO 9000-3 standard and software guideline
- Observance and testing of quality documentation and quality activities

- Documentation of corrective actions for any noncompliances found by means of a corrective action report (CAR)
- Publication of the audit report
- Follow-up by the lead auditor(s) to ensure that the corrective actions have been completed and are effective in achieving the quality goals and objectives of Growth
- The management of customer, vendor, and third-party audits

Schedules

The schedule of internal quality audits is determined by the director of quality assurance. For this purpose, an annual timeline is published by which all areas of the company are audited on an annual basis against the appropriate element(s) of the integrated standard.

The frequency of areas audited is based upon their recent performance and CAR history and the importance of that area within the quality system.

Records

The results of the internal quality audits, recorded on internal quality reports by the auditors, are maintained by the director of quality assurance.

Corrective Action

The corrective action items documented during the internal quality audits are reviewed by the auditor with the assigned company employee directly responsible for the quality activity. Mutually satisfactory date(s) are set for completion of the noted corrective actions.

Follow-Up

The follow-up audits are conducted on the scheduled date(s) for completion of either action items or corrective actions. The results are presented by the auditor to the employees directly responsible for the quality activities audited.

Closure

The action items or corrective actions are closed when the auditor determines that the actions taken are effective. This process may require several iterations.

Presentation

The Director of Quality Assurance presents, by exception, the status of the internal quality audits at the management review meetings.

Escalation

A corrective action not completed according to plan is escalated to the VP & general manager through the management review meetings.

Supplier Audits

The director of quality assurance is also responsible for the management of key vendor audits. The key vendors include assembly houses for outsourced builds; printed circuit board vendors; cable manufacturers; and memory SIMM module suppliers. Supplier CARs (SCARs) are issued, if required, to the vendor for corrective and preventive action.

Critical Production Audit Areas

A comprehensive set of procedures are in place to ensure that all customer and Growth material is properly safeguarded from electrostatic discharge throughout the manufacturing process flow. These include ESD awareness training; wrist, heel strap, and conductive floor wax monitoring; and grounding of floor and bench top mats.

A procedure is used to measure and record solder paste deposition data for sample quantities of all jobs requiring the solder paste screen print process.

A procedure is used to ensure that printed circuit boards meet or exceed customer or Growth cleanliness requirements. An Omega meter is used to set up and audit the circuit-board cleaning process on a daily basis.

Training

All auditors receive internal quality audit training provided by the director of quality assurance. Additional external training is encouraged and sponsored by Growth."

Training Policy

Response

Even though the "Training Program" process document had been completed in fine style, Greg felt uneasy about a solo draft followed by review and called the entire team together to script the training policy. He quickly learned that he wasn't alone in his mixed feelings. Training was not an obvious adventure. It was hard to believe that such a short paragraph in the Standard could involve so much detail.

However, the group put its collective energies to work and essentially banged out the document—which we show below in "Section 18." It turned out that the policy and the process were really very close.

4.18 Training

Growth is to establish and maintain documented procedures for identifying training needs and provide effective training for all personnel.

"Section 18 Training Policy General

Growth's management annually reviews departmental needs and financial resources necessary for Growth's development. The training program is documented in the "Training Program" and "Job Descriptions" documents. The VP & general manager and the director of manufacturing jointly determine the financial budget of the training program.

ISO 9000 training is an ongoing process accomplished through internal quality audits, specific training sessions, company and departmental meetings, and any specific external training as required.

Planning

A working document published by top management is compiled. It is a proprietary document maintained by the VP & general manager and controller. It is made up of annual sales forecasts, departmental budgets, and operational plans for the coming year."

> Personnel performing specific assigned tasks are to be qualified on the basis of appropriate education, training, and/or experience.

"Qualification and Needs Analysis

The departmental managers assess their departmental needs and create job descriptions when needed, which include education level, training, and/or work experience needed to perform specific job functions. In this manner, Growth ensures that all employees are fully qualified for their positions.

Job Descriptions

The job descriptions are created at the time of hiring and maintained and archived in the controller's office. Each job description contains the conditions for education training and/or experience for the employee. The job descriptions are amended as needed upon meeting and interviewing job candidates .

Training

All employees are trained as deemed appropriate by the department manager. Training is in accordance with the individual employee's education, preparation, and previous work experience, as well as Growth's quality commitment. The manager is available to tutor any employee, specifically for any job-related questions. The manager chooses from OJT (on-the-job training), special internal, or external programs.

Specific training programs include classroom work in which all engineering and operational employees receive in-house video training covering all aspects of manufacturing. This includes ESD awareness, OSHA health and safety requirements, proper hand soldering methods and techniques, T/H and SMT process assembly and inspection techniques, component identification, and technical terms and definitions.

Management and professional training (both hardware and software) are provided by OJT or through company-funded courses deemed appropriate by the managers. The training period ceases when the manager deems the employee appropriately trained. Management strives to maintain a quality working environment for all employees at Growth.

Reviews

Employees are reviewed annually by their managers to ensure the highest work quality. Discussions cover work quality, strengths, weaknesses, and potential areas of improvement for the individual and for the department. Previous reviews are maintained and archived. Reviews can be performed on a computer template, or with a written or verbal structure.

Each new employee is reviewed by his or her manager on a time interval decided upon at the date of hire. The results of these reviews are archived in the individual employee's file located in administration. The manager and employee discuss any issues pertaining to the quality of the employee's work. If, in a manager's judgment, an employee requires additional training to improve the quality of his or her work, it will be discussed with the employee and provided for by Growth.

The training programs are provided and run by the department manager."

Appropriate training records of training are to be maintained.

"Records

Hard copy orientation and review records are maintained by the controller, and ongoing training records are maintained in the individual employee files. These are records of all present and previous employees and are archived in each employee's personnel file in administration. Records are maintained for a period of time at the discretion of the individual manager."

SERVICING POLICY

Response

Gary felt very comfortable with a solo first draft because the "Servicing Process" document had been so thorough. The group agreed, and the entire policy was put to bed in under two hours. We show the composition in "Section 19."

> **4.19 SERVICING**
>
> Growth is to establish and maintain documented procedures for performing, verifying, and reporting service activities.

"Section 19 Servicing Policy Procedure

Growth provides to its customers various service functions, which include the following (refer to the documents entitled, "Servicing Process" and "Terms & Conditions"):

- Warranty repair
- Out-of-warranty repair
- Extended service contracts (HW and/or SW)
- On-site warranty support
- Custom hardware or software services
- Field implementation of ECO procedures for software and hardware
- General servicing

Warranty Repair

Growth provides warranty terms and conditions as part of the general sale. Warranty repairs and mandatory ECOs are handled through the "RMA procedure."

Out-of-Warranty Repair

Repairs or ECOs to hardware or software products that are out of warranty are handled through the RMA procedure.

Extended Service Contracts

At the discretion of sales & marketing, Growth offers extended service contracts that may be for hardware or software products.

On-Site Support

Reporting, verification, and tracking of on-site support issues is the responsibility of the customer service manager. Service reports are in the form of memoranda and are completed on-site.

Spare Parts

When appropriate, Growth supplies spare parts to minimize customer downtime. The materials department is responsible for the activity.

Custom Services

Custom services above and beyond Growth's normal servicing policy may be negotiated in an individual contract with the approval of the director or sales & marketing.

Field Implementation of ECO Procedures

The customer service manager oversees the implementation of ECOs that affect hardware or software product in the field. At the discretion of the customer service manager, field exchanges in accord with the RMA procedure may be used to implement ECOs.

General Servicing

The general servicing that Growth provides is defined as verbal and written technical support, historical and analytical information, and other assistance in resolving technical issues."

STATISTICAL TECHNIQUES POLICY

Response

Marlin was somewhat dissatisfied with the process document, "Data & Statistical Analysis Processes," and called the group together to compose the policy.

He felt that their approach had been too superficial and that it was time to take a real close look at statistical process control (SPC) and its possible benefits. He did reach a compromise position out of the group after a morning workout and active discussion. They would start a project to put SPC into production before the year was up.

The group's approach is shown in "Section 20."

> **4.20.1 IDENTIFICATION OF NEED**
>
> Growth is to identify the need for statistical techniques used by the company.

"Section 20 Statistical Techniques Policy Identification of Need

Sampling techniques for incoming raw materials are performed by quality assurance against historically based sampling plans that include 100 percent sampling. Key materials such as PCBs and other integrated circuits are pretested before receipt and/or received with certificates of compliance. The rejection rate on some material—for example, cables—is historically minimal, and a "dock to stock" process is used.

Growth uses data analysis and graphical techniques, such as Pareto charts, to study the results of reliability testing and nonconforming product behavior. A limited program in SPC is also in use and, if successful, will be expanded at the discretion of the director of manufacturing.

In the case of software analysis, field feedback and failure data are collected and maintained by the director of quality assurance in conjunction with the software design manager for review and prioritization by the program managers and the customer service manager."

4.20.2 PROCEDURES

Growth shall establish and maintain documented procedures to implement and control the application of the statistical techniques identified in 4.20.1.

"Procedure

The document that describes Growth's statistical techniques is contained in "Data & Statistical Analysis Processes."

Yield Analysis

Production records the results on a board-by-board basis of first-pass testing. The results are archived in an Excel database and used to plot a first-pass yield. Analysis of yield information is the responsibility of the director of quality assurance."

The Initial Assessment

Status

Fran and Mike grabbed a cup a coffee and sat down to review the ISO landscape. It was now October and the cold, fall rain hinted at how the chill of winter would soon be upon them. Time was running down against their plan; it was time to act.

"So, Mike, where the devil are we—do we call the registrar or not?"

Mike's reply was, "Absolutely, we got 'em right where we want 'em!"

Fran smiled at Mike's somewhat off beat response, but she appreciated his innate enthusiasm. However, it was time for data, which Mike spelled out crisply and clearly. His remarks are summarized as follows:

- All of the Tier I, II, III, and IV documents were through first draft, and many were in final form.
- The internal quality auditors had been trained by Sam, had been certified, and had begun their monthly audits, both internally and at key vendors.
- Several customer audits had occurred, and Growth had performed exceedingly well—to the delight of the whole division and to the great satisfaction of the customers who proceeded to issue new purchase orders.
- The management reviews were underway on a monthly basis and had already caught and resolved several key production issues.
- Corrective and preventive action activity was fully launched and had created a kinship among the employees that was exciting. Team relationships had never been up to this level before.

The Registrar

Fran was delighted at Mike's report and immediately called the four registrars that Sam had suggested. She remembered Sam's admonition that all four companies were acceptable, but that it was a matter of chemistry; she should try to get the companies to send the lead assessor for an interview, if at all possible.

To get interviews with all lead assessors was no easy task. They were very busy people. However, she compromised in several cases and did manage to interview three of the four. The last one just didn't have the time.

Of the three, she picked the registrar who was about midway in pricing, but the decision was really based on what the assessor had to say about his business-oriented interviewing technique and the fact that both he and his associate had many years between them in the electronics industry. He also hit it off very well with other members of the top management team.

His name was Allan D. and his associate was Beth M. Both assessors were certified by either nationally or and internationally known accreditation boards, and the registrar was also accredited by several boards of the same international stature.

The initial assessment would last three days, and it was scheduled for the first week in February, 1998—a few weeks later than planned, but the delay was not an issue.

Preparation

Now it was time for Growth to prepare itself for the February audit.

The Quality Policy Statement was plastered all over Growth, and sessions were held with all employees to make sure that they knew the policy and knew the details in their day-to-day documentation.

Sam and the auditor team held their readiness assessment in mid-November and issued over thirty CARs containing over one hundred observations. No element or area came out of this intensive review unscathed.

The top management team led the charge to correct the necessary problems, and the entire division jumped in with both feet running and showed the kind of spirit and determination that would lead to more success than employees could ever have dreamed possible. Of course, they didn't know that then; they thought they were in really bad shape. These thoughts just stirred them on to clean up their areas even faster.

The IA

December and January seemed to fly by. It seemed that before they could breathe one more time, Allan and Beth were at the door in their snow boots and it was time to either play the fiddle or get off the roof.

Fran and team were sure they were not ready yet, and Fran had fought the inclination to push the IA out a bit. But, they were here and this was it.

At the opening meeting, Fran introduced the division, summarized its history, and welcomed the auditors to the facility. Allan and Beth were as personable and fair-minded as expected, and the assessment went smoothly and was considerably more fun than anticipated. The assessors seemed to mesh beautifully with the staff at all levels and to be genuinely concerned with Growth's potential and ability to achieve a payback for its efforts.

There were several concerns, which would be addressed at the first surveillance visit in six months. One had to do with how preventive actions were being interpreted—not enough credit was being taken; another dealt with three open CARs that had missed their closure promises—which had dates that were too optimistic; and a third had to do with the preventive maintenance procedure that was in the middle of a considerable last-minute revision—however, the PM records were very well kept.

There were three minor findings:

- 4.5 Documentation Control—Three procedures in test, two in assembly, and one in kitting were obsolete and not removed. The procedures were all different.
- 4.8 Product ID—Two parts, a transformer and its winding, were in the transformer build area on the paint bench. Parts were not moved with a job number and were not tagged. The status of the parts was unknown.
- 4.11 Metrology—There were three oscilloscopes, two VTVMs, and a thermometer that were missed in the calibration pick-up, all in the same area of test.

The closing meeting was held in the conference room and Allan, as the lead assessor, praised the preparation of the team and the terrific work that had been accomplished on the documentation. Beth voiced the same feelings and also noted how well each employee interviewed was fully aware of the Quality Policy Statement and knew the most minute details of the work instructions.

Allan then announced that it was his pleasure to recommend Growth for certification to ANSI/ISO/ASQC Q9001-1994, with the condition that the three findings be cleared prior to the certificate issuance. Growth was now free to tell the world that it had been recommended for certification with that one condition.

At that point everyone stood up and cheered, and handshakes were flying at the assessors. A tape was run to capture the spirit of the moment and to show the eighty or so other employees what an initial assessment closing meeting was like. Allan then pointed out that the nonconformances would be cleared through the mail and would probably take a few weeks. The certificate would follow shortly thereafter.

Indeed, that is what happened, and Growth raised its ISO 9001 banner in the first week in March.

AFTERMATH

As 1998 came to a close and the frantic last-minute shipments were made, it was clear that Growth had greatly benefited from the ISO decision.

Sales did level out at around thirteen million, as feared, but bookings in the last quarter foretold of the best start in the history of the division for Q1 1999. This was felt to be a direct result of the increased activities in customer contact and because several large OEMs were far more favorable toward an ISO company. In fact, it was only a short time afterwards when the new contracts were signed.

The big story, however, was the increase in gross margin that resulted from an increased awareness of incoming material performance. RTV and SCAR activity had increased, but first-pass yields had jumped from 67 to 81 percent. The gross margin reached 46 percent—an amazing result.

Fran and the staff also realized that there had been added value in the way the management meetings had contributed directly to the bottom line through their quick response to design and production problems. There just wasn't anything more powerful for decision making than to have a series of charts that clearly indicated the level and scope of profitability and productivity in the division.

They had played it the way Sam suggested, and they would continue to play it again the same way. There was to be no difference between their business and quality goals. They were inseparable. You could actually sense the harmony in this approach as the excited team shared a common theme:

"Growth, Incorporated in 2000!"

Fran began to draft the poster for that slogan.

Appendix A
ISO 9000 Stewardship Summary

Appendix A ISO 9000 Stewardship Summary				
Ref#: **Date(s):**	**Supplier Name/Location:**		**Type: ISO 9001:1994**	
Element ISO 9001	**Activity Covers Assignment of Stewards & Continuous Improvement Managers**	**Stewards by Name**	**C/I Program Managers by Name**	**Comments**
4.1	Management responsibility steward Management representative			
4.2	Quality system steward			
4.3	Contract review steward			
4.4	Design control steward			
4.5	Document & data control steward QA documentation control mngr Engineering documentation mngr			
4.6	Purchasing steward			
4.7	Control of customer supplied product steward			
4.8	Product ID & traceability steward			
4.9	Process control steward Preventive maintenance C/I mngr			
4.10	Inspection and testing steward			
4.11	Control of IM & TE steward Metrology C/I manager			
4.12	Inspection & test status steward			
4.13	Control of nonconforming product steward			

Appendix A

Continued

Ref#: Supplier Name/Location: Type: ISO 9001:1994
Date(s):

Element ISO 9001	Activity Covers Assignment of Stewards & Continuous Improvement Managers	Stewards by Name	C/I Program Managers by Name	Comments
4.14	Corrective & preventive action & customer complaints steward Preventive action C/I manager Customer complaints C/I manager			
4.15	Handl'g, storage, packaging, pres, & del'y steward			
4.16	Control of quality records steward			
4.17	Internal quality audits steward Internal quality audits C/I manager			
4.18	Training steward Training C/I manager			
4.19	Servicing steward			
4.20	Statistical techniques steward			
4.21	Other (safety, security, environ)			

Stewards Manage "channels" of information, that is, policy, process, procedure, and forms, to ensure that the channel is fully documented, implemented, and demonstrating effectiveness in meeting the organization's quality objectives.

C/I Managers Formulate, implement, and effectively manage their specific programs, for example, the preventive action program, the audit program, the training program.

Approved by: **Date:**

Appendix B
ISO 9000:1994
Record Implications

Appendix B ISO 9001:1994 Record Implications			
ISO 9001 Element	**Explicit Records**	**Implicit Records**	**Typical Records Required**
4.1	4.16		▪ Management reviews ▪ Performance graphs (metrics) ▪ Customer satisfaction or dissatisfaction
4.2	4.16		▪ Quality plans ▪ Total quality records ▪ Standards & codes currency
4.3	4.16		▪ RFPs/RFQs ▪ Sales orders ▪ Customer purchase orders ▪ Contract reviews & approvals ▪ Invoices ▪ Customer complaints ▪ Customer specifications
4.4	4.16		▪ Assignment of personnel ▪ Project plans ▪ Project interface documents ▪ Customer design inputs ▪ Design specifications ▪ Design reviews ▪ Design verification results ▪ Design validation results ▪ Design changes

Appendix B Continued			
ISO 9001 Element	**Explicit Records**	**Implicit Records**	**Typical Records Required**
4.5		4.16	▪ Master document control lists ▪ Change notices ▪ Master engineering lists ▪ Retained obsolete documents ▪ Nature of the changes
4.6	4.16		▪ Approved subcontractor lists ▪ Subcontractor monitoring ▪ Purchase orders ▪ Quality requirements ▪ Source inspection records
4.7	4.16		▪ Lost, damaged, unsuitable material reports ▪ Returned material authorizations ▪ Repair records
4.8	4.16		▪ Traceability identifications
4.9	4.16		▪ Production & materials control ▪ Referenced standards & codes ▪ Approval of processes ▪ Approval of equipment ▪ Routers ▪ Drawings ▪ Specifications ▪ Equipment maintenance ▪ Special-process qualifications
4.10	4.16		▪ Receiving inspection ▪ In-process inspection ▪ Final inspection ▪ Vendor certificates of analysis/compliance ▪ Your certificates of compliance ▪ Inspection authority ▪ Pass/fail records ▪ Stamps
4.11	4.16		▪ Test hardware validation ▪ Test software validation ▪ Inspection equipment calibration ▪ Measuring equipment calibration ▪ Test equipment calibration ▪ Master calibration lists
4.12		4.16	▪ Inspection & test records ▪ Waivers/concessions

APPENDIX B **Continued**			
ISO 9001 Element	**Explicit Records**	**Implicit Records**	**Typical Records Required**
4.13	4.16		▪ Rework records ▪ Scrap records ▪ Nonconformance tags ▪ Hold tags ▪ Customer release records
4.14	4.16		▪ Corrective action reports (CARs) ▪ Preventive action reports ▪ Supplier corrective action reports (SCARs) ▪ Customer complaint logs
4.15		4.16	▪ Shipping records ▪ Shelf-life records ▪ ESD records
4.16		4.16	▪ Master records list
4.17	4.16		▪ Audit schedules/assignments ▪ Audit plans/checklists ▪ Audit reports ▪ Vendor audit reports ▪ Third-party audit reports ▪ Customer audit reports
4.18	4.16		▪ Identification of needs ▪ OJT training records ▪ Management training ▪ Training schedules/plans ▪ Qualification records
4.19		4.16	▪ Customer site reports ▪ Warranty records
4.20		4.16	▪ SPC charts ▪ Pareto charts ▪ Sampling plans

Appendix C
Effective "Shalls" per ISO 9001 Element—Expanded Version

Appendix C Effective "Shalls" per ISO 9001:1994 Element—Expanded Version		
ISO 9001 Clause	**Explicit Shalls**	**Shalls by Expansion**
4.1.1	3	11
4.1.2.1	1	18
4.1.2.2	1	6
4.1.2.3	2	5
4.1.3	2	2
4.2.1	3	6
4.2.2	2	5
4.2.3	4	12
4.3.1	1	4
4.3.2	2	4
4.3.3	1	2
4.3.4	1	1
4.4.1	1	4
4.4.2	4	6
4.4.3	1	2
4.4.4	3	5

Appendix C **Continued**		
ISO 9001 Clause	**Explicit Shalls**	**Shalls by Expansion**
4.4.5	3	5
4.4.6	3	4
4.4.7	2	4
4.4.8	1	1
4.4.9	1	4
4.5.1	1	3
4.5.2	3	7
4.5.3	3	4
4.6.1	1	2
4.6.2	2	7
4.6.3	2	5
4.6.4.1	1	2
4.6.4.2	4	4
4.7	2	7
4.8	3	5
4.9	8	14
4.10.1	2	2
4.10.2.1	2	2
4.10.2.2	1	2
4.10.2.3	1	2
4.10.3	2	3
4.10.4	3	5
4.10.5	4	5

Appendix C Continued		
ISO 9001 Clause	**Explicit Shalls**	**Shalls by Expansion**
4.11.1	7	12
4.11.2	2	18
4.12	2	2
4.13.1	2	8
4.13.2	5	6
4.14.1	3	5
4.14.2	1	5
4.14.3	1	7
4.15.1	1	2
4.15.2	1	1
4.15.3	3	3
4.15.4	1	3
4.15.5	1	2
4.15.6	2	2
4.16	7	18
4.17	6	15
4.18	3	6
4.19	1	5
4.20.1	1	6
4.20.2	1	4
Totals	**138**	**320**

Appendix D
Further Examples of Quality Policy Statements

Appendix D Further Examples of Quality Policy Statements		
ISO 9001:1994 Clause	**ISO 9001:1994 "Shall"**	**Quality Policy Statement**
4.3.3 Amendment to Contract	The supplier shall identify how an amendment to a contract is made and correctly transferred to the functions concerned within the supplier's organization.	▪ Excellent's customer service representatives ensure that amendments to contracts are in written form by means of sales orders; reviewed by all affected departments by means of the RFQ report; and then accepted or rejected in written form and returned to the customer.
4.4.8 Design Validation	Design validation shall be performed to ensure that product conforms to defined user needs and/or requirements.	▪ After acceptance by the chief engineer of the final design review and verification testing report, the project engineer contacts the customer and schedules an in-plant validation test with the customer present. ▪ The elevated-temperature electrical and mechanical validation test results are compared to the customer's specifications, and upon acceptance the customer signs off on the final product release form.

APPENDIX D
Continued

ISO 9001:1994 Clause	ISO 9001:1994 "Shall"	Quality Policy Statement
4.5.3 Document and Data Changes	Changes to documents and data shall be reviewed and approved by the same functions/organizations that performed the original review and approval unless specifically designated otherwise.	▪ Document changes are accomplished using the document change form. Changes are reviewed by qualified personnel prior to approval. Revisions made to controlled documents are highlighted to the changes made. Revisions are tracked by the revision number and effective date. ▪ Changes to documentation are approved by those identified by management as the appropriate authority. The document control administrator maintains the authorization lists. Management makes the decision whether or not a change impacts the regulatory status of licensed product.
4.9 g) Process Control	Controlled conditions shall include the following: g) suitable maintenance of equipment to ensure continuing process capacity.	▪ Production equipment and machines are regularly maintained by the maintenance department following the schedules and recommendations provided by their manufacturers, if unavailable, or the recommendations of an appropriate equipment engineer. ▪ Performance and accuracy of the equipment is continuously monitored by the production foremen.
4.10.5 Inspection & Test Records	. . . These records shall show clearly whether the product has passed or failed the inspections and/or tests according to defined acceptance criteria . . .	▪ 100 percent inspection is performed on all boards built on either the prototype or production lines. The inspections are performed by quality control personnel, and pass/fail data are recorded on the traveler by means of QC stamps, and the records are maintained in the quality assurance laboratory files.

APPENDIX D

Continued

ISO 9001:1994 Clause	ISO 9001:1994 "Shall"	Quality Policy Statement
4.18 Training	The supplier shall establish and maintain documented procedures for identifying training needs and provide for the training of all personnel performing activities affecting quality.	▪ The personnel department or representative at each of Excellent's plants is responsible for coordinating all training programs and for maintaining all records pertaining to training. This process is documented in SOP TR18-01. ▪ New employees at each plant receive, as a minimum, instruction on plant safety, the ISO 9000 quality system overview, skills instructions, and basic statistical concepts. ▪ The personnel department or representative at each plant surveys each department at least annually to identify OJT and cross-training needs. The information is reported to the general manager, who prepares the training plans and issues the training schedules to each department manager. ▪ Each department manager creates the required OJT exercise to fulfill those needs and assigns an appropriately skilled trainer or trainers.

Appendix E
Checklist for ISO 9001:1994 Element 4.5 Document and Data Control Quality Manual Requirements

APPENDIX E Checklist for ISO 9001:1994 Element 4.5 Document and Data Control Quality Manual Requirements					
Ref#: Date(s):	Supplier Name/Location:			Type: ISO 9001:1994	
ITEM	Mandatory Activities to be Covered in the Quality Manual	1st Draft	1st Edit	Final Edit	Release Date
1.0	Define the Tier II process doc (procedure)				
1.1	Controlled document attributes -Quality manual -Tier II processes or procedures -Tier III work instructions -Tier IV forms				
1.2	Hard copy vs electronic media -Access limits -Backup protocols -Read/Write protocols				
1.3	How documents are created				
1.4	Review & approval protocols -QA documents -Eng'g documents -Documents of external origin				
1.5	Mix (manuals vs individual docs) -Tier II master list -Tier III master list -Tier IV master list				

Appendix E **Continued**					
Ref#: **Date(s):**	**Supplier Name/Location:**			**Type: ISO 9001:1994**	
ITEM	**Mandatory Activities to be Covered in the Quality Manual**	**1st Draft**	**1st Edit**	**Final Edit**	**Release Date**
1.6	Method of distribution/removal				
1.7	Control method (central vs LAM) -QA documents -Engineering documents -Documents of external origin				
1.8	Obsolete document protocols				
1.8.1	Normal obsolete removal				
1.8.2	Retained for info/legal purposes				
1.9	Revision protocols				
1.9.1	Review and approval				
1.9.2	Nature of change control				
1.9.3	Pertinent background protocol				

Definitions: LAM = Local area manager

APPENDIX F
AN EXAMPLE OF GROWTH'S PROCESS FLOW-CHARTING PROTOCOL

Hub Document for Quality Audits

Quality Audit Processes

Text Supplement for the Quality Audit Flow Charts

The flow chart entitled "Internal & External Audit Process," File XAUDIT.ACL, describes the quality assurance audit process in detail and introduces the safety audit process that is found in the flow chart entitled "Safety Audit Process," File SAFETY.ACL. In all cases, the flow chart takes precedence over this text.

The senior lead auditor (Sr. LA) manages both programs.

Lead Auditor Training

Each lead auditor completes the training program shown in the workbook entitled "Systems Lead Auditor Training Manual." This program may be run by either an outside qualified source or the senior lead auditor.

An outside qualified source must either be individually certified under a national or international certification schema or be an employee of an accredited training program, for example, a RAB certified lead assessor, or RAB accredited training program.

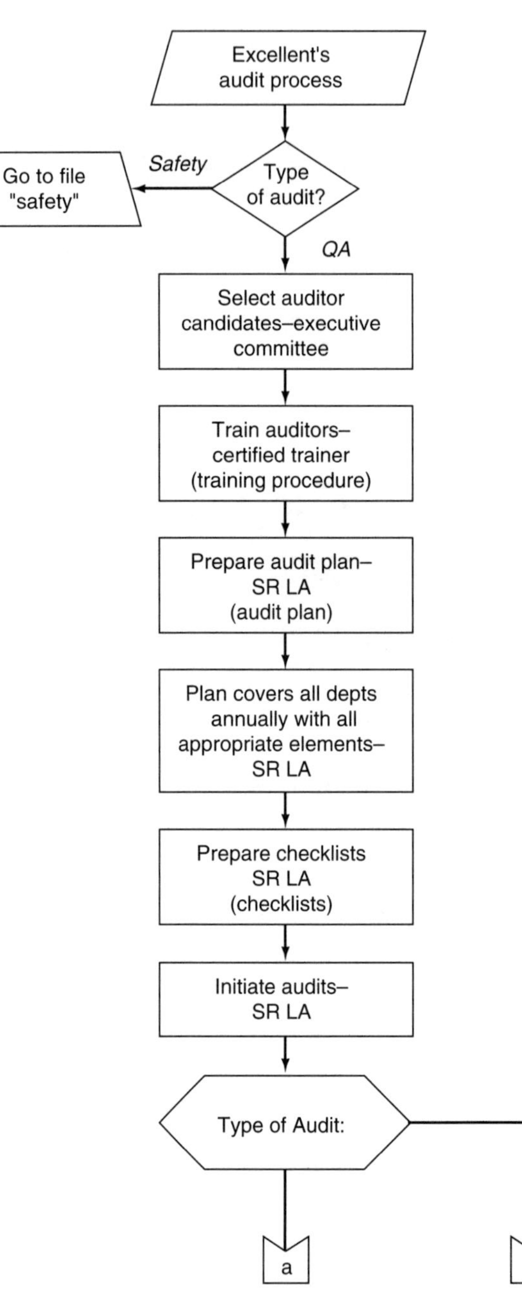

Figure Appendix F

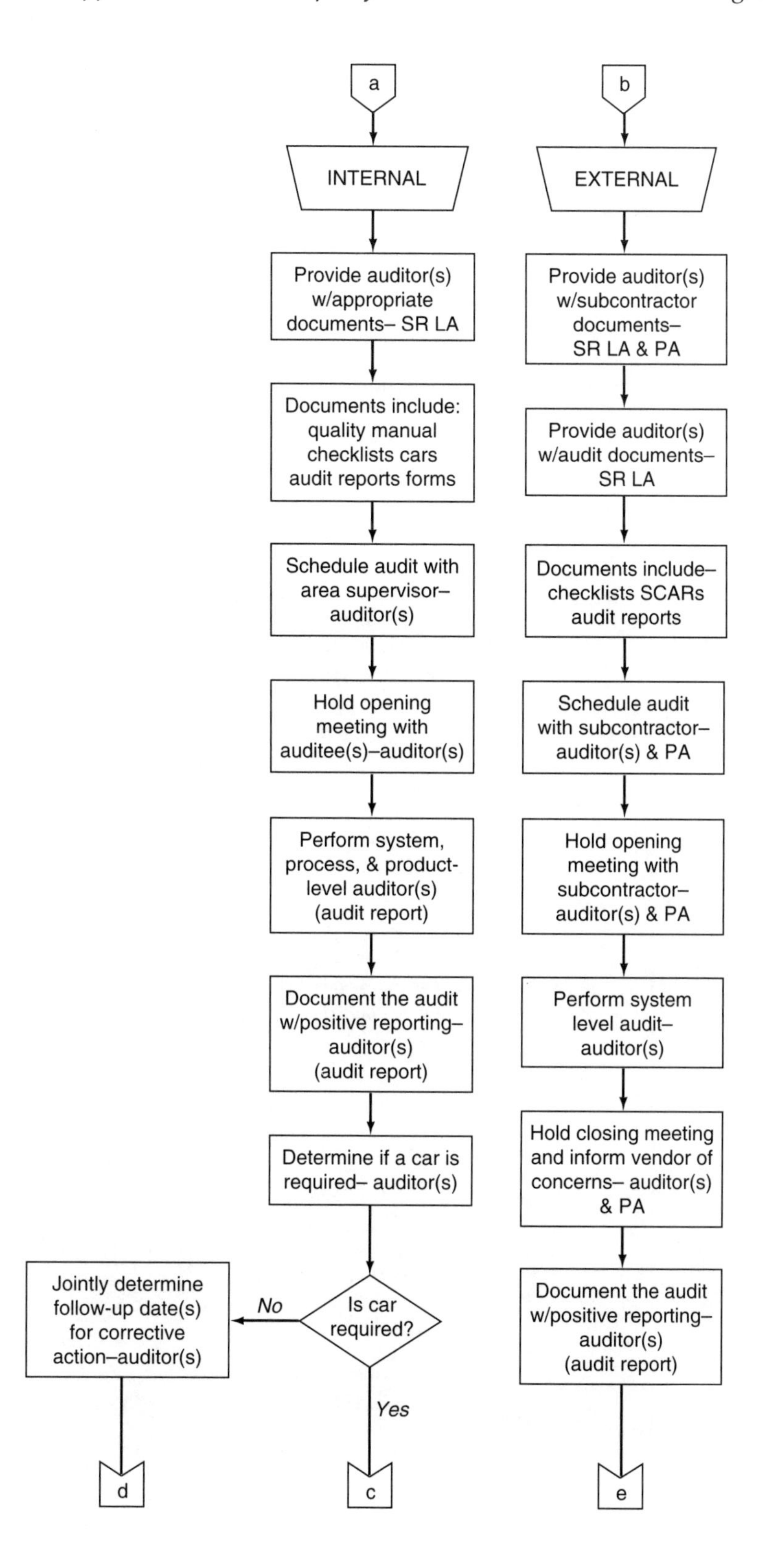
a
INTERNAL
Provide auditor(s) w/appropriate documents– SR LA
Documents include: quality manual checklists cars audit reports forms
Schedule audit with area supervisor– auditor(s)
Hold opening meeting with auditee(s)–auditor(s)
Perform system, process, & product-level auditor(s) (audit report)
Document the audit w/positive reporting– auditor(s) (audit report)
Determine if a car is required– auditor(s)
Is car required?
No
Yes
Jointly determine follow-up date(s) for corrective action–auditor(s)
d
c
b
EXTERNAL
Provide auditor(s) w/subcontractor documents– SR LA & PA
Provide auditor(s) w/audit documents– SR LA
Documents include– checklists SCARs audit reports
Schedule audit with subcontractor– auditor(s) & PA
Hold opening meeting with subcontractor– auditor(s) & PA
Perform system level audit– auditor(s)
Hold closing meeting and inform vendor of concerns– auditor(s) & PA
Document the audit w/positive reporting– auditor(s) (audit report)
e

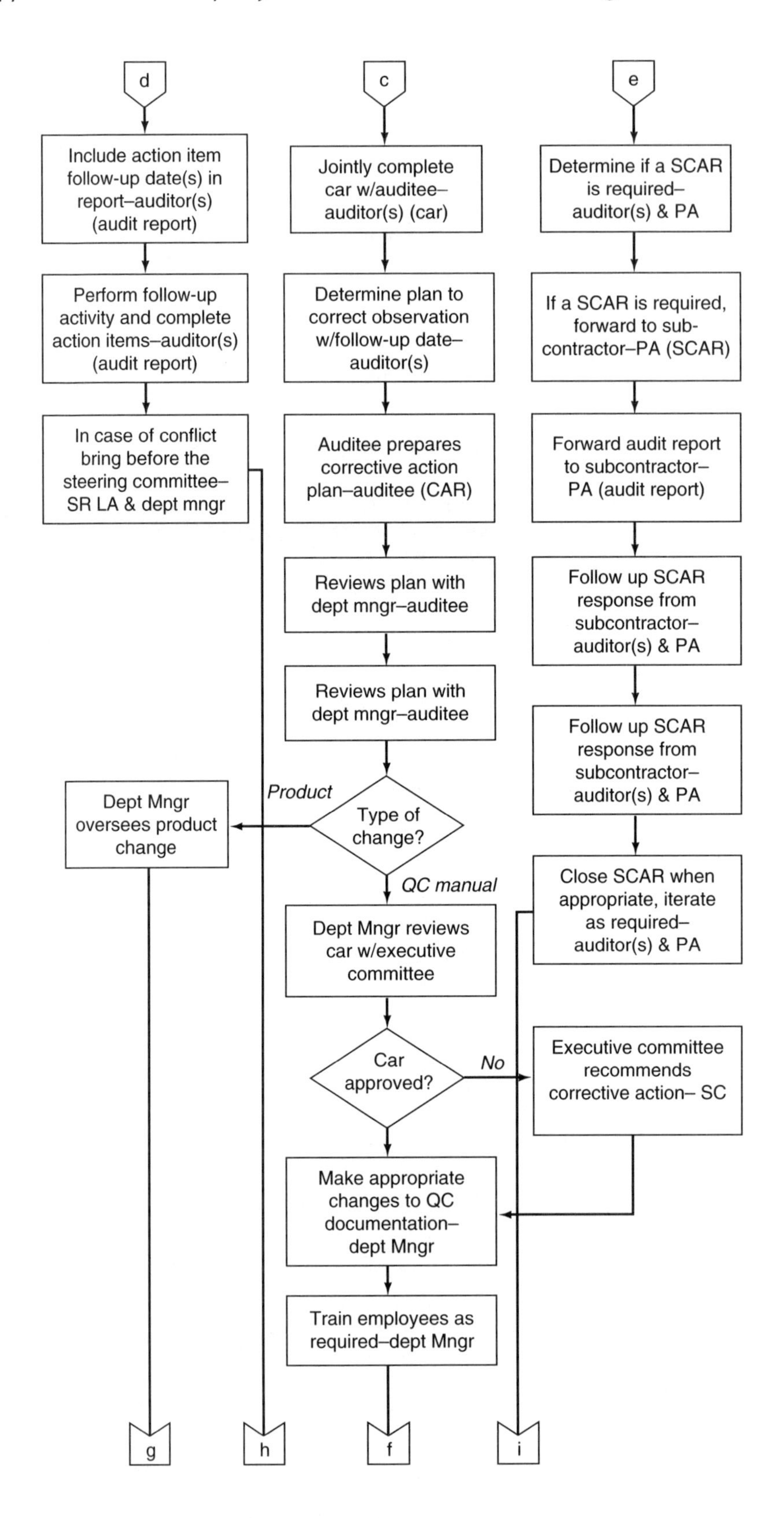
d
c
e
Include action item follow-up date(s) in report–auditor(s) (audit report)
Jointly complete car w/auditee–auditor(s) (car)
Determine if a SCAR is required–auditor(s) & PA
Perform follow-up activity and complete action items–auditor(s) (audit report)
Determine plan to correct observation w/follow-up date–auditor(s)
If a SCAR is required, forward to sub-contractor–PA (SCAR)
In case of conflict bring before the steering committee–SR LA & dept mngr
Auditee prepares corrective action plan–auditee (CAR)
Forward audit report to subcontractor–PA (audit report)
Reviews plan with dept mngr–auditee
Follow up SCAR response from subcontractor–auditor(s) & PA
Reviews plan with dept mngr–auditee
Follow up SCAR response from subcontractor–auditor(s) & PA
Dept Mngr oversees product change
Product
Type of change?
QC manual
Close SCAR when appropriate, iterate as required–auditor(s) & PA
Dept Mngr reviews car w/executive committee
Car approved?
No
Executive committee recommends corrective action– SC
Make appropriate changes to QC documentation–dept Mngr
Train employees as required–dept Mngr
g
h
f
i

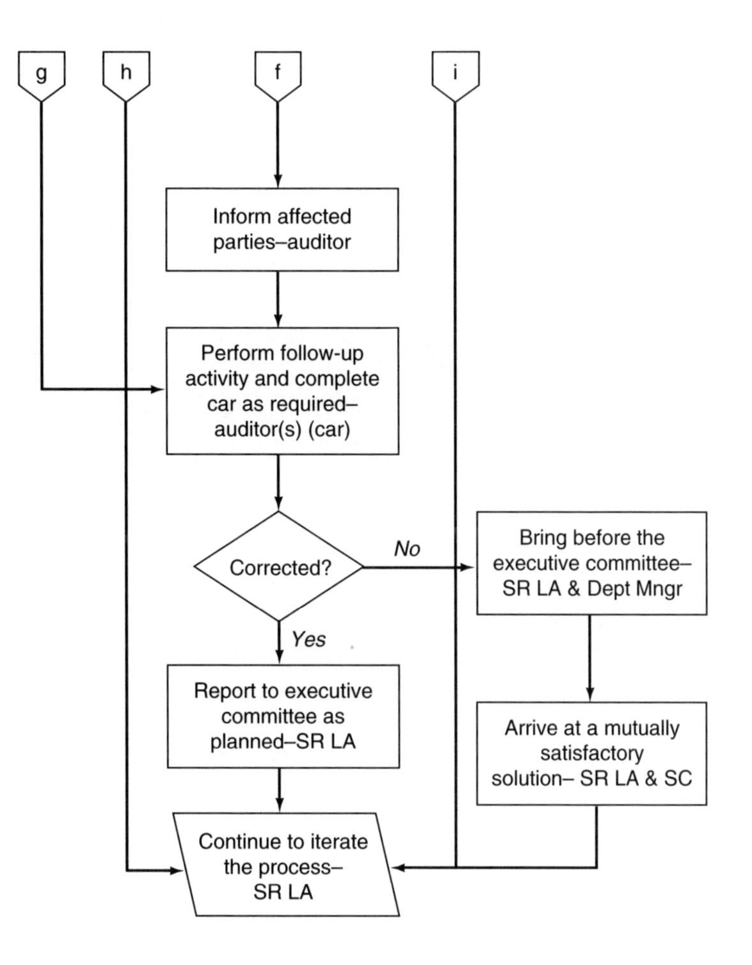
g
h
f
i
Inform affected parties–auditor
Perform follow-up activity and complete car as required–auditor(s) (car)
Corrected?
No
Yes
Bring before the executive committee–SR LA & Dept Mngr
Arrive at a mutually satisfactory solution– SR LA & SC
Report to executive committee as planned–SR LA
Continue to iterate the process–SR LA

Appendix G
Bibliography

AIAG (1995). *Quality System Requirements, QS-9000.* Detroit, MI. Ph: 810-358-3003.

ANSI/ISO/ASQC, American National Standard Series by the ANSI/ISO/ASQC (Fall/Winter 1997), ASQC Quality Press Publications Catalog, Section on Standards.

- ANSI/ISO/ASQC Q9001-1994, Quality Systems—Model for Quality Assurance in Design, Development, Production, Installation, and Servicing
- ANSI/ISO/ASQC Q9000-3-1991, Quality Management and Quality Assurance Standards—Guidelines for the Application of ANSI/ISO/ASQC 9001 to the Development, Supply and Maintenance of Software
- ANSI/ISO/ASQC Q9004-1-1994, Quality Management and Quality System Elements—Guidelines
- ANSI/ISO/ASQC A8402-1994, Quality Management and Quality Assurance—Vocabulary
- ANSI/ISO/ASQC Q10011-1994, Guidelines for Auditing Quality Systems
- ANSI/ISO/ASQC Q10013-1995, Guidelines for Quality Manuals
- ANSI/ISO 14001-1996, Environmental management systems—Specification with guidance for use.

ASQ, American Society for Quality, Milwaukee, WI.

- *Quality Management Journal*
- *Journal of Quality Technology*
- *Quality Engineering*

Bates, Jefferson D. (1985). *Writing with Precision,* Washington, DC: Acropolis Books LTD. 2009.

Brumm, Eugenia K. (1995) *Managing Records for ISO Compliance.* Milwaukee, WI: ASQ Quality Press. (ISBN 0-87389-312-3), 6 × 9 hardcover.

Crews, Frederick (1974). *The Random House Handbook.* NY: Random House.

Campanella, Jack, ed. (1990), *Principles of Quality Costs.* 2nd Ed. Milwaukee, WI: ASQC Quality Press.

Deming, Edwards W. (1986). *Out of the Crisis,* Cambridge, MA: MIT Press.

FDA (1996). *Quality System Regulation,* Part VII, Department of Health and Human Services, Food and Drug Administration, *21 CFR Part 820, Current Good Manufacturing Practices (CGMP):* Final Rule, October 7, 1996, published by the SAM Group, Stat-A-Matrix, Edison, NJ, Sec. 820.181 Device master record.

Hammer, Michael and James Champy. (1993), *Reengineering the Corporation.* New York: Harper Business, HarperCollins Publishers.

Horn, Robert E. (1989). *Mapping Hypertext.* Wiltham, MA: The Lexington Institute, Information Mapping, Inc.

IMI (1994). *Demystifying ISO 9000.* 2nd ed. Waltham, MA: Information Mapping, Incorporated.

Ingalls, Karyn E. (1997). *Demystifying CGMPs.* Waltham, MA: Information Mapping, Incorporated.

IRWIN. (1996). *Registered Company Directory, North America.* November. Burr Ridge, IL: IRWIN Professional Publishing.

ISO. (1996). *ISO Standards Compendium. 6th ed. ISO 9000 Quality Management.* Geneva: International Organization for Standardization (ISBN 92-67-10225-7).

Juran, J. M. (1992). *Juran on Quality by Design.* New York: The Free Press.

Kuhn, Thomas S. (1970). *The Structure of Scientific Revolutions.* 2nd ed., enlarged, Vol. II, No. 2. Chicago: Foundations of the Unity of Science, University of Chicago.

Medical Device & Diagnostic Industry. (1997). Santa monica, CA Canon Communications, LLC. Monthly publication.

NIST U.S. Department of Commerce, National Institute of Standards & Technologies. Ph: 301-975-2002.

Peach, Robert W., ed. (1994). *ISO 9000 Handbook.* 2nd ed. Fairfax, VA: McGraw-Hill Companies.

Porter, Michael E. (1985). *Competitive Advantage.* New York: The Free Press.

Quality. (1997). Carol Stream, IL: Hitchcock Publishing, Co. Monthly publication.

Quality Progress. (1997). Milwaukee, WI: American Society for Quality. Published monthly.

Quality Systems Update. (1997). Fairfax, VA: McGraw-Hill Companies.

Randall, Richard C. (1995), *Randall's Practical Guide to ISO 9000.* Reading, MA: Addison Wesley.

Tiratto, Joseph. *Registrar Accreditation, The ISO 9000 Handbook.*, 2nd ed. Fairfax, VA: The McGraw-Hill Companies. See also another McGraw-Hill publication, *ISO 9000 Registered Company Directory, North America.*

Index